THE ECONOMIC CASE FOR PALESTINE

THE ECONOMIC CASE FOR PALESTINE

ELIAS H. TUMA and HAIM DARIN-DRABKIN

CROOM HELM LONDON

© 1978 Elias H. Tuma and Haim Darin-Drabkin
Croom Helm Ltd, 2-10 St John's Road, London SW11

British Library Cataloguing in Publication Data

Tuma, Elias Hanna
    The economic case for Palestine.
    1. Palestine – Economic conditions
    I. Title   II. Darin-Drabkin, Haim
    330.9'5694'05   HC497.P2

    ISBN 0-85664-559-1

Printed and bound in Great Britain by
REDWOOD BURN LIMITED
Trowbridge & Esher

CONTENTS

List of Figures and Tables
Preface

1.	Introduction	13
2.	Viability of Nation States	17
3.	The Case for a Palestinian State	34
4.	The People and the Land	47
5.	The Economy	60
6.	The Viability Prospects I – Sector Contributions	70
7.	The Viability Prospects II – The Overall Picture	89
8.	Regional Integration	103
9.	Summary and Conclusions	114
	Appendix	118
	Bibliography	120
	Index	124

LIST OF FIGURES AND TABLES

Figures

2.1	National Independence	21
3.1	Projected Boundaries to the Two-State Solution	45

Tables

3.1	Basic Population Estimates for the Palestine State	44
4.1	Population of the West Bank and Gaza 1968 and 1975	47
4.2	Pattern of Settlement, 1975	48
4.3	Housing Distribution per Family	50
4.4	Employment Structure, 1968 and 1975	52
4.5	Distribution of Population According to the Districts and Corresponding Areas	54
4.6	Cultivated Area According to Categories and Regions	55
4.7	Urban Population and Space in Municipal Administered Lands According to the 1967 Population Census	56
4.8	Land Use Distribution Estimates in the West Bank	59
5.1	Land Use Distribution by Crop: West Bank	61
5.2	Crop Distribution of the Agricultural Produce in 1975	62
5.3	Change in Real Value of Agricultural Output and Real Income of Farmers (per cent change)	63
5.4	National Accounts of the West Bank and Gaza Strip in 1975	68
6.1	Distribution of Palestinian University Graduates as of 1969	72
6.2	Use of Urban Space	79
6.3	Projected Urban Land Allocation	80
6.4	Proposed Industries and Area per Worker in each Branch	82
6.5	Human Settlement Costs	86
6.6	Estimates of Infrastructure Costs for One Million People	87
7.1	Population, Labour and Output in the Transition Period	91
7.2	Employment and Income Structure in the Transition Period	92
7.3	Additional Investment Needed for Housing and Employment Creation	93
7.4	Sources of Capital	95
8.1	Composition of Exports and Imports of Jordan, Israel and the Palestinian State Territory	107
8.2	Per Capita Income in West Bank, Gaza, Israel	109

A1   The Output and Investment per Employee According to
        Commodities   118
A2   List of Products and Findings of Import Forecasts for
        Investigated Markets   119

PREFACE

This study marks the completion of three years of co-operation between an Israeli Jew and a Palestinian Arab. Our joint effort arose spontaneously, as we discovered that we both search for the same thing, namely peace, security and self-determination for both Jews and the Arabs who claim Palestine as their home. This goal we still share, but we also share a deep interest in approaching the problem in a rational and objective way. We have tried to look at the facts and potentialities and evaluate them dispassionately to the extent possible. We have stayed away from the political issues and from the moral questions of justice, historical claims and where the blame lies. To get to the point, we have assumed hypothetical conditions which seem to us highly probable as bases for a state of Palestine. Should these hypothetical conditions be modified, the conclusions reached will have to be reconsidered, but the framework of the study will continue to serve as a model for reaching new conclusions, and hopefully will lead to other and more specialized studies.

Our thanks to all the people who have been instrumental to bring this study to completion. Special thanks to Anastasios Papathanasis who helped in the research and to Pam Rush who did an excellent job in typing the manuscript.

31 July 1977   EHT and HDD

To all Arabs and Jews who wish to live in peace and harmony with their neighbours, and to the people who make it possible for them to do so.

1 INTRODUCTION

Much has been written about the Arab-Israeli conflict, the prospects for peace, or war, and the eventual establishment of a Palestinian state, side by side with the state of Israel. The emphasis, however, has been on the political process of such eventualities. Our objective complements these previous efforts. We are concerned with the economic aspects of these various solutions. In particular, we are concerned with the economic feasibility of a state of Palestine: assuming the policy makers have agreed on a solution, will the resulting state of Palestine be economically viable? What minimum conditions must be met for it to survive and prosper? What size population can it support, what boundaries should it have, and what period of time must elapse before the full potential and viability of such a state can be realized?

These are some of the questions we shall address. It is our intention to provide at least a skeletal outline or an economic blueprint for the creation of a state of Palestine west of the Jordan River. Given the prospective, though still undetermined, features of a state of Palestine, we may approach the question by exploring the economic feasibility of small nation-states in general. Against this background we may then be able to assess the feasibility and prospects of an economically viable Palestinian state. This study is based on the following hypotheses:

1. Small states may require economic conditions that are different from those necessary and/or sufficient for the viability of large states. The optimum size of the state is relative in the sense that what is small or large depends on what we compare the given unit with. In this context, the relative minimum size is taken to reflect the ability of the state to depend on its own resources for its economic survival. The ability of the state to secure the necessary inputs and to dispose of its output in a way that contributes to the welfare of its citizens would be fundamental to its size viability. It is true that autarky is hardly ever possible, but relative potential self-sufficiency is basic to economic independence and the ability of an economy to survive under pressure from other nations. A large state would presumably have land, labour and other resources to sustain its production processes. A small state, in contrast, would rarely have an adequate variety and quantity of the various necessary resources to sustain itself, without trading with other nations. Therefore, those states that are not large enough to be relatively

self-sufficient must have certain economic or political advantages to offset the size handicap, such as alliances, strategic location, or certain other unique qualities which may be used in a bargaining situation.

2. Self-sufficiency may not be ideal or even desirable in a world of international trade and interdependence. What may be lost because of small size or the inability to be self-sufficient may be compensated for through trade, international co-operation, and a certain degree of planning. A state's ability to export would offset shortages in local resources by gaining access to imports. Aid from other countries may be another method, though such a means tends to create instability because of the inherent dependence of such a relationship. Planning, on the other hand, may assure viability by creating a closed economy, by restricting imports to the mere essentials, and by scaling down the expectations of the population and thus creating contentment. In other words, viability becomes a function of expectations which in turn would be tailored to fit the capacity of the economy. As in the case of dependence on alliances, dependence on the manipulation of expectations may be helpful in the short run; in the long run it tends to breed discontent and instability.

3. Economic viability is necessary but not sufficient for national survival: it must be supported by political and social viability. Regardless of the size of the state, economic endowment must be used in harmony with the political institutions on which the state is based. They must also contribute to social harmony within society. In other words, economic viability must be considered within the framework of political and social viability. For example, poorly used resources can hardly sustain viability; inequality of income and wealth distribution may undermine the contribution of resource endowment to economic viability of the state.[1]

4. Economic viability is a function of endowment, scale of production, technology and expectations. It follows from the above that viability must be supported by a set of contributing factors. Endowment with resources, including labour would only determine the potential for viability. The scale of production and the level of technology would determine the costs of production, competitiveness on the market, and the productivity and income of labour. These factors indicate whether an economy is capable of achieving viability; however, realized viability is a function of expectations which determine the level of incentives and contentment. Put in these terms, viability then may be independent of the size of nations, measured in terms of area, population, or total output. It is, rather, a function of

productivity, competitiveness and expectations.

5. An Arab State of Palestine west of the Jordan River may be economically viable and fully harmonious with an Israeli State. Though the boundaries are still unspecified, it is our hypothesis that the geographical area previously known as Palestine is capable of sustaining a Jewish state and an Arab state. Such viability, must, however, depend on internal as well as external resources. Trade, aid and planning must play major roles in the early life of the Arab state to render its potential viability realizable.

6. To be economically viable, it is necessary for the national state to have well defined and legitimate boundaries, such that only minimum expenditure would have to be devoted to defence. It is reasonable to assume that establishment of such boundaries would also enhance economic resources, such as tourism, and productive expenditure in the form of investment. It would also enhance economic and trade relations with other states within the region and outside it. Indeed, we may suggest that unless this condition can be satisfied, economic viability may be unrealizable.

7. On the basis of our investigations, we suggest that a state of Palestine west of the Jordan has the potential to achieve high enough productivity and income favourably comparable with those of other countries in the region. Such an achievement, however, depends closely on the will and determination of the Palestinians and the co-operation of other parties with them. The human resources that make the Palestinian people must play the most important role in achieving viability.

These hypotheses form the framework of our study. The next chapter will deal with hypotheses one through four, by analyzing the meaning of and general conditions for economic viability in various sizes and types of nation-states or economic units. The analysis will explore the interaction between the various resource endowments. We shall evaluate the impact of changing the technology on the relations between factors of production in modern society. The analysis will be illustrated by comparison and contrast between countries from various regions and time periods, given the resource endowment, scale of production, technology and trade effects on the respective degrees of viability.

Chapter 3 will introduce the case for a Palestinian state and specify the alternative boundaries for such a state. These alternatives will be assessed according to the principles of viability discussed in the preceding chapter, the historical bases for considering them, and the political and social feasibility of implementation. This chapter will conclude by selecting the alternative set of boundaries that seems most feasible according to our analysis.

Chapter 4 will be a case study of the people and the land that will constitute a Palestinian state. We shall give a brief survey of the area, the climate and the people as a point of departure. Chapter 5 discusses the economy as it has been and will estimate the potentialities of the economy. Chapters 6 and 7 deal with the prospects of viability, on a sectoral basis at first, and then in an aggregate summary form. The prospects are evaluated by comparing potential achievements with the required results to achieve viability in terms of income, employment and investment in quantitative terms. We also assess the subjective conditions of viability or the human question.

Chapter 8 considers the Palestinian state within the regional context and explores analytically the various implications of making the Palestinian economy a part of the larger region of the Middle East, or at least of its immediate surroundings. We are skeptical of the suggestions that the Palestinian State should be integrated with the neighbouring countries.

Chapter 9 will summarize the findings and note the areas that require deeper study to understand viability in general and to appreciate the economic viability of the Palestinian state in particular.

Note

1. This statement is subject to debate, depending on the potential harmony between the objectives of the state and the economic philosophy of its people.

2 VIABILITY OF NATION STATES

Meaning of Viability

Viability, according to Webster, means the capability of 'growing and developing'. Politically the concept has been used to mean the ability of a party to annihilate its opponent, or its ability to resist being annihilated.[1] This concept means the ability to survive relative to other parties it has contact with. The question of economic growth enters only in as much as it becomes necessary for survival as in the situation where the economy of the opponent country has in the meantime grown enough to put the former party at a disadvantage. Used in an economic context, a society is considered viable 'if its economic characteristics permit it to experience sustained growth and rising welfare per capita'.[2]

These different meanings of viability either show total relativism by tying viability to the conditions of other parties, as in the case of conflict, or show concern only for domestic conditions, as in the case of sustained growth and rising welfare; the former approach tends to ignore what happens within the country in micro terms, and the latter pays no attention to what happens elsewhere even within the region in which the country happens to be. These interpretations also fail to indicate the critical limits or minimum conditions to be viable: what does the ability to resist being annihilated mean? Would survival through an alliance with a third party mean viability of the state, even though it implies dependence on that party? Boulding describes this as conditional viability, which tends to be unstable. Or, what does sustained growth or rising economic welfare mean? Would a rise of income or of the growth rate of .001 per cent be sufficient to assume or sustain viability? Would an annual rise in economic welfare by a fraction of one per cent imply viability? These interpretations do not address themselves to the issues, presumably because the meaning and interpretation of these concepts must be relative to the immediate environment. To be able to grow and develop, for example, the economy must have assured resources, a market, and must not be too dependent on others.[3] To remain viable, a country must grow and develop at least at a rate equal to those achieved in the immediate region; otherwise, the economy would fall behind, lose its relative strength, and become vulnerable to domination by its opponents.

17

Recent studies have advanced the thesis that each country would do well to develop at its own pace, on the basis of its own resources, and by its own technology. Though that may seem ideally appropriate, in reality no country exists in a vacuum. Expectations and taste evolve in the context of the experience the people are faced with. Hence, viable growth can hardly be divorced from the growth experience of the region in which a country happens to be.[4]

It may be difficult to give a single and universally accepted definition. Therefore, an operational definition of economic viability may be *the ability of the economy to attain and sustain an economic structure and level of per capita income comparable to those of other nations in the region with comparable resources.* To avoid vulnerability, or unviability because of dependence on another party, viability may require an additional condition: the ability to use a country's own resources directly or through trade to achieve national targets. That is, while an economy may be short of certain basic resources, to be viable it ought to have surpluses of other resources, goods, or goodwill to trade for the goods that are in short supply. This means that a viable economy must be able to achieve its targets without having to depend on perpetual aid from the outside, though trade and temporary aid are fully consistent with viability.

According to this definition, viability seems to be an asset that may not be accessible to many nations, given the unequal distribution of resources, and the widespread dependence of certain nations on others. Yet, in actuality, the achievement of viability may be less difficult than seems to be implied, as will become clear when we explore the bases on which viability depends.

Prerequisites of Viability

Viability depends on subjective and objective conditions or criteria, which must exist simultaneously for viability to become a reality. The subjective criteria include the will to develop, a positive or ambitious level of expectations, and behaviour patterns that sustain and realize the expectations. In other words, viability requires commitment, confidence and the belief that the expectations can be realized, and action in consistency with these commitments. To illustrate, a commitment may be reflected in the pronouncements made by the public authorities committing the state to promote development. The expectations may be reflected in the targets established in advance by the policy makers to shoot for. Implementation represents action. However, implementation would lead to viability only if it is carried out at a rate and a level of efficiency that would realize the objectives within a specified period or time horizon. In other words, plans and targets must be accompanied by the allocation of the necessary resources and action to implement these plans.

The behaviour patterns which relate to viability are critical, such as attitudes towards work, sacrifice and persistence. The targets may be high or low and the expectations may be very ambitious or not at all. As long as the performance can lead to realization of the objectives, we may speak of subjective viability as achievable. Interpreted in terms of economic activity, this criterion describes the labour force motivation, the spirit of enterprise, and the discipline and organization of the production process at all levels. In his assessment of the growth of small nations, Simon Kuznets considered the impact of endowment and concluded that endowment without the ability to utilize it would not suffice. The crucial variables must be 'sought in the nation's social and economic institutions'. Kuznets identifies the following as crucial features of successful small nations: (1) internal homogeneity, since it facilitates co-ordination, which often characterizes small nations; (2) a certain degree of egalitarianism; and (3) the ability to adjust to the changes in the international trade market, or the ability to show a capacity for 'social invention'.[5]

We should also consider the impact of culture and tradition on behaviour to assess the degree to which viability may be augmented, rather than hindered by the prevalent cultural and traditional values. In particular, viability requires that the economic target be consistent with the national sense of unity. To the extent that viability is being used as an aggregate concept, it requires that the citizens of the state have a sense of unity which enhances the commitment to achieve viability. This is what may be called a sense of national identity.

The question of identity may be raised in another way: what makes the unit a nation and what significance does the nation concept have for economic viability? Politically, the nation concept implies political authority, and a unit of action in decision-making. Authority and decision-making, however, are related to various cultural and institutional facts such as language, religion, education, etc. To act as a nation, the unit must, therefore, feature a certain degree of homogeneity, unity of purpose, and continuity of action within its boundaries. Discontinuity demonstrates the external boundaries of a nation, identifies it as an internally unified entity — it reorganizes its assumed or real identity. For purposes of economic analysis, a nation may be identified according to the international framework, both because of differences from other countries in economic and other institutions, and because of the internal unity of the market implied in economic nationalism.[6] A nation may be small but enjoy a large market if it reduces the discontinuity and promotes integration with other nations. If such integration is concluded willingly, the national identity may be strengthened; if imposed, however, the national identity may be compromised; more on this in Chapter 8.

The subjective criteria discussed so far may be summarized as an identity confidence. The success of a nation-state depends on the establishment of a national identity which the citizens of the state would acknowledge, promote and defend. Both the senses of identity and of confidence are necessary for economic viability since both the political and economic factors enter into the creation and sustenance of the nation-state. The necessary sacrifices through reduced private consumption, a reallocation of resources in favour of national causes, and the initiative and dedication to hard work on behalf of the group are reflections of the commitment to the national identity. The identity criterion is especially significant in the case of new states, small states, and the not-very-well endowed. The new state must be able to make a breakthrough in the international community to be accepted; self-confidence is the first step in that direction. Small states must be able to establish their credibility to be able to play national roles which are often disproportionately larger than the size impact of the state itself. It is hardly significant for a state with 200,000 people to be concerned with the cold war between two super powers, except possibly as an intermediary and the carrier of good will. Yet, the small state must enter international diplomacy with a high degree of confidence in its identity to be effective or even to be heard at all. Finally, states that are not well endowed may have problems in the establishment of identity confidence because of their dependence on others and their limited ability to make the necessary contributions in the relations between nations. This contradiction is depicted clearly in Figure 2.1. Probably the confidence of identity is most significant and indispensable in the case of the state that combines all three handicaps: newness, smallness, and low endowment. The subjective criteria in such a case become critical in the assessment of economic viability and the viability of the nation state as a whole.

The objective criteria are the physical conditions which characterize the economy and relate to the production process. Generally these are the factors which constitute or influence the forces of supply and demand in the aggregate. Viability requires that the forces of supply and demand be in relative balance at a level of per capita output which

Figure 2.1: National Independence

'*American plant, Russian equipment,*
German assembly, Czech parts
—all the fruit of our
good old neutralist know-how.'

Source: Adapted from Leopold Kohr, *Development Without Aid,* The Merlin Press, Llandybie, Carmarthen, 1973. Reproduced by permission of *Punch*.

is favourably comparable with per capita output in the region and consistent with the national expectations. As expectations rise (they never go down) or as the level of per capita output in the region changes, it should be feasible for the process of production to change accordingly. If supply fails to respond, demand will go unsatisfied and output will lag behind. If this lag lasts for a relatively long period the economic viability of the system will be in jeopardy. The signs will be reflected in the low per capita income, in the negative balance of payments, in a falling exchange rate, and possibly in a political crisis in the state.

If, on the other hand, the forces of demand fail to respond, the producers' confidence will be shaken, output will decline, and a crisis could develop. While there may be an apparent balance between supply and demand at a lower level of per capita output — under full employment — that balance is inconsistent with the economic viability of the state: it represents unemployment, under-utilization of resources, inability to compete on the international market, and a potential crisis of confidence. Again, if the crisis lasts for a relatively long time, the economic viability of the state would be in question.

The forces of supply, or the factors of production, are of two kinds, those which are irreproducible (natural) and those which are reproducible at a relatively high cost. The natural or irreproducible resources are the critical ones which must exist at least at a minimum for the state to exist. Given the population, these include land, water, the climate and a certain level of native know-how. These resources can hardly be compensated for, short of population movement or emigration. If mere space does not exist or the land is of such quality that it cannot be reconditioned for productive purposes at a reasonable cost, or if there is no water to be had at a reasonable cost, then development of the economy is virtually impossible. On the other hand, reproducible capital such as raw material and equipment, are possible to acquire from other countries. Whether it would be economical to do so would depend on the availability of complementary inputs and the comparative advantage a country enjoys. The reproducible capital, important as it is, cannot be a major factor in determining viability; capital availability may determine the rate at which development will proceed, but not whether it will proceed or not. In contrast, irreproducible capital is a critical factor and can determine whether or not viability is feasible. Land space, water and the climate are probably the only critical factors, although the quality of each of these would determine the ease with which development can advance and the relative cost of producing an income.

It is important at this point to dispel certain misconceptions about viability. It has been common in the literature to stress the significance of reproducible capital as a major factor. Yet empirical evidence shows that the role of capital is at best ambiguous. Historically, the availability of capital has had only limited impact on economic development, and the lack of capital was often easily compensated for by domestic policy or international agreements.

A difficulty arises when we try to estimate the necessary capital for viability. Not only is it virtually impossible to make such an assessment, but often it is not even possible to arrive at a standard and usable definition of capital. Because of these ambiguities, which can be clarified only in the specific context of the given economy, and because capital can be acquired by various means, we rule out capital as a critical or determining factor in economic viability.[7] On this basis we may judge potential viability by assessing the ability to maintain a balance between population and the critical resources.

To achieve and maintain a viable economy, a certain land/labour ratio or a resource/population ratio, must be satisfied as a minimum; and a certain quantity of water for each of the domestic, industrial and agricultural uses must be secured. These ratios vary by the type of economy, the composition of the output, and the size of the population.

The relevant level of technology may be considered a separate factor of production, or it may be regarded as a characteristic of the labour force. If the latter, manpower would then have to be differentiated according to skill and entrepreneurship. Economic viability would depend on the quality of labour, the level of literacy, skill and professional training, as well as on the spirit of initiative and risk taking. While it is not possible to specify a critical minimum percentage of labour in each skill, it is essential that a major component of the labour force be native to the state and/or permanent residents, fully familiar with the culture, and have a wider interest in the country than this economic or business activity. The native labour force should have a cadre of experts in all the basic fields of endeavour, with emphasis on the areas of domestic specialization. No state can be viable if it is dependent to a large extent on foreign expertise, unless such dependence is partial, voluntary and mutual. Though an economy may continue to function for a long time at a primitive level with little expertise and low literacy, such an economy can hardly attain viability especially if its population continues to grow. Its apparent viability may be in question once contact with the outside creates a demonstration effect and new expectations, or if change and development in the region are rapid enough to cause the given state economy to lag behind. Thus, the degrees of literacy, expertise and entrepreneurship necessary for viability are variable, according to the dynamic conditions of the specific economy. Viability requires that no serious manpower bottle-necks develop and last for any length of time.

The economy of the state, or the production process, depends on the availability of land as a natural resource. Land plays a major role as a place of residence, as a site for industrial enterprise, and as the source of food, minerals and other agricultural and industrial raw material. The importance of the land factor, as well as of water and minerals, depends on the size of the population or the labour force. Factor proportionality in the production process determines the level of technology, the ability to maintain full employment, and the capacity of the economy to retain the level of self-sufficiency assumed at the beginning of the production process. Let us take a few illustrations.

Scenario 1: The population increases, land is available to engage and support the population increments; assuming no water or capital shortage, output per capita may be maintained and the standard of living may be sustained. If the land supply is still relatively abundant, more capital-intensive cultivation may be practised, and the yield per unit of land and/or labour raised. The economy would remain viable in these circumstances.

Scenario 2: The population increases, but no land is available to absorb the increment. The result may be a shift to industry, a decline in the standard of living, or emigration. If a shift to non-agricultural occupations is not possible, one of the other two alternatives would be the obvious outlet. The only way to avoid emigration or a lower standard of living may be to enforce a radical population policy that would restrict population growth and maintain a balance between land and labour.

Scenario 3: Population increases but neither of the above policies seems feasible. The only other alternative would be territorial expansion, which could spell imperialism for the strong and defeat for the weak, neither of which would be a viable long-run solution. Accordingly, the achievement of viability on the supply side requires a balance between land and labour and technology over time and in changing situations.

An important question is how much land per person is necessary for economic viability; and what kind of land. These questions cannot be answered in the abstract, since the appropriate factor proportions will depend on expected income, the level of technology, and available alternatives for employment, and the role of agriculture in the economy. The amount of land necessary for viability also depends on the demand for land for other uses such as parks, recreation and the infrastructure. General estimates may be made, but it is more important to remember that viability depends on maintaining a balance. It is equally essential to note that the land should be good enough for cultivation to sustain the population dependent on it. In other words, rather than worry about the land area for agriculture, we should be more concerned with the cropping capacity of the land and its potential output.

The viability of the supply in the economy depends also on the infrastructure, including roads and other means of communication, and the economic and legal institutions that facilitate and protect the flow of inputs into the production process. Included in this set of pre-requisites is a financial system which facilitates capital mobility and creates incentives for the suppliers to respond to the changes in demand.

The supply question is only one side of the economic system. Without demand production becomes economically irrelevant. Unless there is demand for the product, it would be irrational to go on producing — at least in the economic sense. Demand is a function of population — for subsistence, of income, of government expenditure, and of investment expenditure to produce for the domestic and foreign markets. This means that as population increases, demand would also increase. However, unless purchasing power is also available, such demand would have no significance. Therefore, viability requires that the economy be able to generate enough purchasing power and expenditure to attract supply and absorb it.

A balance between the forces of supply and demand would reflect the responsiveness of the supply and demand forces to each other and the possibility of attaining viability. It is more important, however, that dynamic viability be attainable or that a change in demand would stimulate sufficient supply to satisfy it. It also means that an under- or over-supply would instigate price or policy changes to generate demand to absorb the supply change and thus restore a balance at a 'high' level of achievement. Dynamic viability is especially relevant in view of population growth, resource depletion or the discovery of new endowments and techniques.

Traditionally we would expect the market mechanism to bring the forces of supply and demand into balance with each other. This may still be feasible if the economy develops at its own pace, in normal circumstances, and if a large enough market may be expected. But if the economy is small, is newly created, and its market institutions are relatively backward, complications may be expected. Two patterns of development may be foreseen as means to attain viability. The first is to depend on the market, develop slowly, rely on domestic resources, exploit primary production possibilities, promote isolationism if necessary, and utilize largely domestic skills. Viability in the long run should be attainable in this model if no external interferences occur and natural population adjustments are allowed to take place.

However, a more rapid and somewhat guaranteed approach is to plan the economy and balance the forces of supply and demand, adopt a target growth rate and co-ordinate the various sectors accordingly, secure aid and grants from other countries, emphasize industrial production, use foreign skills, and become integrated in the larger economic environment. Or, the policy makers may choose to plan and reduce the interdependence with the outside and depend primarily on domestic resources and skills.[8] Isolationism or an open economy are choices that may be consistent with the market mechanism as well as with planning, with corresponding consequences and responsibilities.

It depends on the economics of the region and the domestic pressures for attaining viability whether forces other than the market should be operative. For example, if a natural balance between land and labour does not prevail, it may be unavoidable to take steps to bring about such a balance. Similarly, while isolationism may be inconceivable for a small state, it should be obvious that integration in the larger community may have to be brought about gradually as the economy approaches a relatively advanced or competitive level of production and efficiency. Until then, viability may have to be avowed and protected, and industrialization consciously promoted. It is equally significant that the economy should be able to produce a large proportion of its food requirements. If this does not happen through the market forces, the plan must bring it about by direct or indirect methods.

Given these viability requirements, is it possible for a small nation-state to achieve viability? Is there a minimum size necessary to achieve it? Are there minimum rates of growth, levels of technology, or rates of industrialization that form the threshold for viability? These questions may be answered only by assuming certain conditions and expectations.

Models of sustained development and growth have been formulated, although no model of economic viability has been proposed.[9]

It is easy to begin by invoking the take-off model of W.W. Rostow,[10] according to which an economy reaches the take-off stage when all the preconditions for take-off have been satisfied. These include a stable government, a fairly comprehensive network of infrastructure, an entrepreneurial class that is willing to invest and take risk, and an interest in applying science and technology to the economy. The take-off stage, however, is the critical step toward development into a modern economy. In this stage investment rises above 10 per cent annually, productivity increases, and industry expands. But achieving a take-off does not guarantee viability. The take-off may be achieved by external or transitory influences such as foreign aid, war or temporary alliances. When these influences subside, the economy may or may not have become independent and viable, and may stagnate or decline as a result.

It is significant that Rostow's model emphasizes functions and results, regardless of the size of the economy or of the nation. Apparently the size makes little difference as long as the basic functions are performed and the transition is made possible. This assessment is consistent with other findings in the literature, as well as with the eclectic theories of development that are in existence. In his interesting study of the size of nations, Nadim G. Khalaf has found the role of the foreign market to be independent of the size of the nation. He also found: 'First, that there is no significant association between size and rates of growth of GNP, nor between size and increases in per capita GNP... Second, that there is no significant association between population size and per capita GNP.'[11]

These findings suggest that size as such may have little relevance. However, it is the combination of characteristics or set of criteria that may account for development or viability. It is possible that the larger the size of the nation, the higher the probability that these criteria may be satisfied.

In his study of small countries, William G. Demas took the conditions for development and sustained growth as criteria. Development in his analysis means structural transformation of the economy which includes: reduction of dualism, removal of surplus labour, elimination of subsistence production, increase of industrial production in response to demand, and more diversification. On the basis of Chenery's survey of the patterns of development and his own analysis, Demas concludes that the problems faced by small countries are similar to those faced by large countries. However,

4. In a very small economy there are limits to import substitution, and transformation usually involves the export of manufactured goods unless (a) the country has primary or resource exports in high demand, or (b) there is not heavy population pressure in agriculture.

5. A small country which has to produce manufactures finds this difficult for a number of reasons, the most crucial being the existence of economies of scale in the production of most kinds of manufactured goods.

6. This implies an export drive coupled with, where possible, varying forms of economic co-operation between such countries.

7. A dynamic theory of economic integration is an essential element in the economic development of small countries . . .

9. Apart from the smallness of the domestic market, there are other disadvantages in small size, the most important being the difficulty of preventing the export of savings and the sacrifice of domestic employment to balance of payments equilibrium in an open economy and the difficulty of developing a broadly based capital market.

10. Among the economic advantages of small size are 'the importance of being unimportant' in external commercial policy, more unified national markets, greater flexibility, and perhaps greater potential social cohesion.[12]

Demas applies his analysis to the Caribbean. It seems evident to him that small economies may achieve transformation, though with some difficulty. The mechanism, however, involves a certain measure of planning and a certain level of integration to compensate for smallness, but it can be done.

Demas' findings are consistent with a study, sponsored by the government of Trinidad, of the economics of nationhood, according to which economic integration of the Caribbean into a federation is considered essential. The reason is that 'all evidence seems to point to a minimum size of between two millions and three millions being necessary for the "take off" into sustained economic growth'. But side by side with federalism and economic integration, 'the government of

Trinidad and Tobago considers that the economics of nationhood ... demand a government absolutely and completely independent'. A Federal Government, in attempting the profoundly difficult task of laying the foundation of a national economy, must have complete command of all its material and other resources, including its perspective for the future. It is equally necessary that the nations and national and international organizations with whom it deals should have for it that confidence and respect which can come only from their recognition of its power to handle and assume responsibility for all its affairs. Any concession whatever on the question of absolute independence is an acknowledgement of weakness and instability.[13]

Studies of Belgium and Switzerland support most of the above observations, although they emphasize the features of these 'efficient' economies more than how these features have come about. Vinelle, for instance, emphasizes the ability of the small industry in small countries to overcome market limitations through export or a stimulus for expansion, but not how a country like Belgium manages to do that.[14]

Similarly, a study of Switzerland lists as explanations of the high level of economic efficiency the freedom from war, internal political stability, a high degree of industrialization, a high level of exports, subsidiaries in foreign countries, a large stock of real capital, a high rate of investment, a high quality labour force, a high level of research and entrepreneurial energy, and high flexibility and adaptability of the economy. It would be most helpful if we could find out how Switzerland came to possess these features.[15]

It is apparent by now that endowment may be neither a necessary nor a sufficient factor for viability. As long as compensating advantages prevail, material resources may be acquired and utilized. The cases of Israel, Switzerland, Belgium and Lebanon are good illustrations of how the lack of material endowment could be compensated for. On the other hand, many well-endowed countries in Latin America, Asia and Africa have been in a rather difficult situation trying to raise their income level, self-sufficiency and efficient utilization of their endowments.

Furthermore, endowment with resources may have little significance in advancing development and viability unless the resources are utilized to increase income and raise productivity in the economy, either directly through resource processing, or indirectly by utilizing the revenue from the sale of the resources to create a processing or manufacturing industry. In other words, viability requires that the economy should develop beyond the production and sale of raw material and primary products. The ability to co-ordinate the various factors of production into a production process and co-ordinate the forces of demand to stimulate supply and consume the output is basic to economic viability. None of the above prerequisites, however, should be dependent on the size of the economy *per se*. In fact the small size may be an advantage, since it may be easier to manage, while it forms less of a threat to others in the international market than the large economy. The significance of size, however, is more conspicuous in as much as a small area would have a lower probability of being endowed with many factors of production than the large area. Hence, the small economy tends to be less capable of self-sufficiency and more dependent on others compared with the large economy.

Whether a 'relatively' small economy can be viable or not may be suggested by the recent proliferation of small and mini national economies around the world and as members of the United Nations. Is it possible that viability is so certain that the world body and the national leaders of the respective new states show little concern for the size of the nation?

The proliferation of small states is political first and social and economic second. Given the emerging tendency toward independence and self assertion, both the national leaders and the world body find themselves caught in the drama of creating a national identity for people who had sought independence for long, with little immediate concern for economic viability. It is evident that economic viability would mean little if the national identity is not asserted first. Accordingly, small nations arise, secure recognition, and then proceed to establish economic viability.

The political bias, however, is only one possible explanation. It is equally significant that the new small nation-states have been emerging in a more congenial environment than the above discussion has implied. The new nations are aware of at least three major forces that render viability feasible. First, the advances in technology, which are easily accessible to all nations, make the factor proportions highly flexible and reconcilable with what is available. To take a simple example, the technological feasibility of greatly increasing the cropping capacity of a given land area is sufficient to make limitations of the land endowment relatively insignificant. Similarly, the advances in transportation have made the lack of a given resource relatively easy to overcome by importing that resource from the outside. In other words, the idea of viability through self-sufficiency has become less expected than previously.

Second, there has been a wide acceptance of the idea of interdependence with other nations, both economically and politically. Given the doctrine of comparative advantage and specialization, it has been fully accepted that no country should seek to be viable on its own. Specialization in accordance with comparative advantage would both increase total welfare and secure viability through a system of constructive interdependence. The market, bilateral agreements, and alliances would make such interdependence safe for the trading nations so that viability through others becomes acceptable.[16]

Finally, a major factor contributing to the proliferation of small nations, regardless of the prospects of economic viability through domestic resources, has been the tendency to favour regional integration among nations with geographic proximity and political affinity. Integration within the region has been considered the last cure to the problem of smallness, whether it is the smallness of the market, the limited endowment of the economic unit, or the disproportionality of factors in the economy. The prospects for the small economy to become a part of the larger regional unit without seriously compromising its political and national identity has probably been the most important factor contributing to the emergence of small, not well endowed, and apparently unviable states. Integration raises their hopes that they would be able to replace unviability with viability, and dependence with interdependence, and thus their national aspirations would be realized.

Given the definition of economic viability and the prerequisites which have been identified, we need to determine the behaviour pattern or policy which would most probably generate viability. It is probably safe to suggest that no single pattern can be considered ideal for all situations. The appropriate policy or pattern of economic behaviour depends on at least four variables: the time horizon, the ideology, the point of departure and the environment.

The impact of time is directly related to the target rate of achieving the income objective or viability status. An economy that suffers from high population pressure, inferiority complex, or other such handicaps must adopt a policy of rapid growth to avoid a Malthusian trap, defeatism, and a self-generated stagnation. In contrast, a country that can choose its pace of growth has no such fears. To a large extent the time horizon is a function of the gap between the present condition and the target.

The ideology is critical to the extent to which policy makers often adopt policies that may be economically irrational, although they may be politically and socially rational. Probably the best illustration is the ease or difficulty with which policy makers resort to planning or insist on depending on the market mechanism, regardless of the economic pressures facing them for rapid change. Ideology in many ways determines the readiness of the people to make sacrifices, live austerely for a while, or adopt egalitarian policies, all of which may have an impact on the rate at which viability can be achieved.

Both of the above factors are related to the starting point toward the objective. If still at an early stage of development the economy would be under pressure to and can achieve a relatively high rate of development, and to approach economic viability at a relatively rapid pace. Such an economy tends to have unused capacities and a lot of catching up to do. But as incomes go up, it becomes harder to grow fast and less urgent to do so. Thus, the time horizon begins to expand and ideology may even become less dominating.[17]

Finally, the pattern of behaviour must be explained within the environmental context, both the internal conditions and forces of the economy, and the external conditions surrounding the economy and society. The policy toward viability depends on whether the economy is backed by a stable mature political system, whether a well-developed educational system exists, and whether the social and cultural values are conducive to expansive economic behaviour. It is equally significant whether the economy operates within a war or a peace environment, or an environment of co-operation or cut-throat competition. If we take all these factors together it becomes clear that while economic viability may be feasible, it has to be achieved and the pattern of behaviour must be tailored to fit the conditions of the economy under consideration to make that achievement possible.

To summarize, economic viability is a function of a complex of characteristics and behaviour patterns which enable the economy to cope with demographic change, to generate and sustain a standard of living progressively approaching the expected level, and to withstand outside economic pressure for a reasonable period of time. The levels of technology and education, the banking institutions, the transportation system, and the stability of government are all among the characteristics of viability. Though no critical set of conditions can be proposed for each of these characteristics, the impact of their interaction on the level of income will be the guide to their assessment and modifications in the environment of the total economy.

Our interest in the economic viability of small states is a result of our concern with the Arab-Israeli conflict. More specifically, we are concerned with the economic feasibility of establishing a viable Palestinian state, side by side with the state of Israel, west of the Jordan River, as a solution to the conflict. Given the fact that such a state must be relatively small, apparently not well endowed, we pose the question: Can such a state be economically viable? The answer in the context of the above discussion must be conditionally affirmative. Viability is possible but it must be achieved. It is not inherent in the prospective state, but it is achievable. The rest of the study will be a 'test' of these hypotheses to the extent possible.

Notes

1. Kenneth Boulding, *Conflict and Defense,* Harper Torchbook, 1963, p.58.
2. Vivian Bull, *The West Bank — Is It Viable?,* Lexington Books, 1976, p.143.
3. This is what I have elsewhere called one-sided dependence, see my 'International Interdependency and World Welfare' in a forthcoming book, *The New Economic Issues,* Nake M. Kamrany (ed.) (EHT).
4. For an excellent exposition of the self-contained view of development see Leopold Kohr, *Development Without Aid. The Translucent Society,* Llandybie, Carmarthenshire: The Merlin Press, 1973, esp. Chs. 1, 4 and 6.
5. 'Economic Growth of Small Nations' in E.A.G. Robinson (ed.), *Economic Consequences of the Size of Nations,* NY: St Martin's Press, 1960, pp.27-32.
6. I. Svennilson, 'The Concept of the Nation and Its Relevance to Economic Analysis'; and C.N. Vakil and P.R. Brahmanada, 'The Problems of Developing Countries', in E.A.G. Robinson (ed.), *Economic Consequences.*
7. A most interesting discussion of capital requirements may be found in Solomon Fabricant, 'Perspective on the Capital Requirements Question' in Eli Shapiro and W.L. White (eds.), *Capital for Productivity and Jobs,* Prentice-Hall, 1977, pp.27-49.
8. This pattern has much in common with the self-generated, domestically-oriented development approach proposed by Kohr, though the mechanism he advocates has no application in this context; Leopold Kohr, *Development Without Aid,* esp. Ch.4.
9. Models approximating economic viability will be reviewed briefly below.
10. W.W. Rostow, *Stages of Economic Growth,* Cambridge University Press, 1960.
11. Nadim G. Khalaf, *Economic Implications of the Size of Nations,* Leiden: E.J. Brill, 1971, p.231.
12. *The Economic Development in Small Countries with Special Reference to the Caribbean,* Montreal: McGill University Press, 1965, pp.19-20, 90-91.
13. Trinidad, Office of the Premier and Ministry of Finance, *The Economics of Nationhood,* Trinidad, 1959, pp.3, 16.
14. Duquesne de La Vinelle, 'Study of the Efficiency of a Small Nation: Belgium' in E.A.G. Robinson (ed.), *Economic Consequences.*
15. W.A. Johr and F. Kneschaurek, 'A Study of the Efficiency of a Small Nation: Switzerland', in E.A.G. Robinson (ed.), ibid. pp.54-77.
16. The probability that the small nation would actually become more dependent, rather than interdependent, has only recently become an issue, though it has come about in the context of Third World debates against dependence on the developed countries.
17. For historical tendencies see H. Chenery and M. Syroquin, *Patterns of Development, 1950-1970,* Oxford University Press, 1975, Ch.2; a comparison of China and India illustrates these points very well.

3 THE CASE FOR A PALESTINIAN STATE

In Search of a Relevant Solution

The creation of a Palestinian State side by side with the state of Israel has been a recurrent proposal as a solution of the Arab-Israeli conflict. Such proposals have usually been based on humanitarian, legalistic and socio-political considerations. A State of Palestine would allow self-determination, independence and the re-establishment of a homeland for the dislocated Palestinians. Though the proponents of these proposals may have been fully aware of the difficulties to be faced by the new state, they have not discussed these issues openly, especially the economic problems. Three major problems stand out among those facing the Palestinian State as visualized in the literature: the State of Palestine will be relatively small, completely new, and founded on bitter memories of decades of hatred and war. Therefore, even to begin to approach economic viability, the Palestinian State must be able to overcome the problems of smallness, newness, and not-so-friendly an environment, in addition to the apparent poverty of natural endowment, regardless what territorial dimensions the new state may take. Nevertheless, the proponents of the independent state solution have enough confidence to continue to promote that proposal in spite of these difficulties.

Most of the previous proposals have been oriented toward solving the political conflict between the Arabs and the Israelis. Only a few attempts have been made to assess the economic prospects of such a state, but none in sufficient detail, as warranted by the complexity of the problem.[1]

Our concern stems from the urgency of the problem and from our identities as an Israeli and an Arab. Our co-operative effort should, we hope, bring a new perspective into the picture. We are concerned that the solutions so far advanced have left many economic questions untreated. We are convinced that a political solution will be found, and that it will involve the creation of a Palestinian State. When that happens, we would like to make sure that the economic problems are known and ways of dealing with them have been mapped. As we explore the economic aspects of the new state, we hope to focus attention on what may be needed to assure the economic viability of the new state. For example, what is the population capacity of the land

34

area proposed for the state? What resources will be available for exploitation? What are the endowments of the prospective state, its labour force, market facilities, institutions and infrastructure? These are some of the questions to be addressed below, keeping in mind the viability prerequisites discussed above. Two main points will be treated in the rest of this chapter: the viability prospects according to the studies already completed and the dimensions of the state that seem most consistent with our joint viewpoint. To deal with the first, we shall survey various territorial proposals that have been made, the alternatives that seem possible, and then specify the dimensions that seem most consistent with our joint assessment and conception of a viable state of Palestine.

Viability Prospects in the Literature

It is not too surprising that little research has been done on the prospects of the economic viability of a Palestinian State. The emphasis has been on the political prospects, on the assumption that once the political solution has been found, the economic problems would be easily overcome. This assumption is most apparent in the literature of the Palestinians and the other Arab parties who are involved in the conflict. While large numbers of publications have been produced, almost all of them deal with the legal, social, political and emotional aspects of the conflict. The lack of emphasis on or research in the economics of the area is not fully a matter of neglect. On the one hand, economic viability is a relative concept. Compared with the present economic situation of the Palestinians, the creation of a State of Palestine can hardly be anything but an improvement. To be rid of economic 'exploitation' by the occupiers, to be relieved of the refugee camp life, to be able to pursue economic activity freely as a citizen of the State of Palestine must seem an immense achievement and a good argument that the state would be viable. In other words, the subjective conditions for viability may then be satisfied, when looked at from an Arab Palestinian perspective.

On the other hand, the apparent lack of emphasis on the economic problems of the prospective state is rather consistent with the ideological commitment of the Palestinians, especially the Palestine Liberation Organization (PLO), to the establishment of a unified, secular, democratic state in the whole of Palestine. According to this logic, the creation of a unified state in the whole of Palestine would satisfy the objective conditions for viability, as well as the subjective ones. The total resources, population, market and trade advantages would now be at least as good as they have been before unification. If Israel alone may be considered economically viable, a unified Palestine should be even more viable. Therefore it should be obvious that the issue of economic viability may be dismissed as an obstacle in the way of a solution.

Probably the best representation of this approach is contained in the writings of Yusuf Sayigh as an economist and former head of the planning department of the PLO. In an important article written in 1970, Sayigh explained the Palestinian concept of a peaceful solution and compared it with the Israeli concept of settlement. He took the whole of Palestine as the given territory, with the Jewish and Arab populations as its human resources. In addition to the Jewish people in Palestine, Sayigh drew attention to the 60,000 Palestinian Arabs who held university degrees in 1970. He also noted that 1,500 Palestinians were professors, physicians, engineers and scientists in Europe, Canada and the United States. These, in addition to those graduated since then, would be regarded as an invaluable human resource to sustain the viability of the proposed state. Details, however, or resource complements were not given or apparently taken into consideration. Whether other studies by the PLO for internal use have been prepared is not known to us.[2]

The economic prospects of the West Bank and Gaza have been the subject of study as *territories* under Israeli rule, and under specified assumptions.[3] The assumptions of Ben Shahar *et al* include: East Jerusalem is separated from the West Bank; free trade with the labour mobility into Israel are either non-existent, or they are total; trade with the East Bank is either non-existent or totally free; capital is not a constraint; skill is a limiting factor in industry and manufacturing; water is a limiting factor in agriculture. Given these assumptions and the fact that military spending in the territories will continue, low and high estimates of growth are generated for the decade of 1968-78, divided in two sub-periods. Except for the impact of the 1973 war, the basic conditions have remained the same, which justifies taking a serious look at the results of the study.

Output projections range from a rate of growth of 5 to 13 per cent a year in the first five years, and from 7.8 to 10.2 per cent in the second five years ending in 1978. Employment, however, would increase by much less and unemployment would not be eliminated. Of particular significance is the conclusion that the high estimates of growth are contingent upon free trade and labour mobility into Israel. Full employment will not be achieved, nor will unemployment be reduced unless workers are allowed to seek employment in Israel. The highest rates of growth are projected to be in construction, especially if work in Israel is permitted. The annual rate of growth of construction, however, would still be high if work in Israel is stopped, as long as new housing for settling the refugees is undertaken.

Assuming a positive development policy by the authorities, it is estimated that agriculture may absorb about 70,000 workers, if trade with the East Bank is allowed and water supply is about doubled within ten years. Manufacturing is considered least promising, given the assumed scarcity of skill and technical expertise. The services, on the other hand, would continue to be the home of under-employment and disguised unemployment for many years to come. The overall prospects do not seem too bright, even if contact with Israel is maintained. As concluded by the authors, sustained growth is possible through 'exogenous development supported by outside supply not only of capital, but also of entrepreneurship and markets for the developing industry. In short, there is a direction for action, but it would be foolish to assume that it would not require a long time.'[4]

Recent estimates fail to support these projections. Given the relatively small change in agricultural investment, employment in agriculture in the West Bank and Gaza changed very little between 1968 and 1975 — it actually declined from 45,000 to 44,000. Industrial employment increased from 18,000 to 20,000, while construction employment declined from 13,000 to 10,000. The major increase was in public and community service, from 18,000 to 25,000 and in trade from 36,000 to 40,000. Employment in Israel, however, increased most, from 5,000 in 1968 to 66,000 in 1975.[5]

The study of Ben Shahar *et al.* has only limited relevance to our analysis for several reasons: it does not consider the independent status of a Palestinian State and the potential supply of skilled manpower contingent upon its creation; it does not allow for the possible inclusion of East Jerusalem and for the role of tourism in the economy of the state; it also under-estimates the burden of increasing population by leaving out the possibility of having to absorb refugees who will be returning from the East Bank and from other Arab countries, as would be expected if a Palestinian State is established.

The impact of the occupation on the development of the economy has been negative in a number of ways. The industrialization may have been hampered by

the shortage of local entrepreneurs, skilled labour, know-how, and plants and equipment; and the lack of a proper banking system and money market ... to these should be added the political uncertainty and competition on the part of Israeli industry. Despite the official policy of encouragement, very little Israeli entrepreneurship and capital have been invested in the areas' industry, because of constraints stemming from both the political situation and the economic risks involved.[6]

A different interpretation of the occupation impact has been reached by Jamil Hilal with respect to the West Bank. Hilal sees the Israeli policy as imperialism which exploits Arab labour and resources and renders the West Bank economy totally dependent on and complementary to the economy of Israel, or on the East Bank through the policy of 'Open Bridges'. As a result, the West Bank economy has been unable to make progress toward an industrial independent economy.[7]

The prospects of a Palestinian State are treated more directly but more briefly by Richard Ward who draws attention to the fact that an independent state would be different from an occupied territory. Given the destructive impact of the 1967 war, Ward suggests that 'to maintain a viable economic state in the long term the GNP of the West Bank must maintain at least its pre-1967 war rate of growth of six to eight per cent a year'.[8] This means that all industrial enterprises would be restored, the markets regained, and the labour lost because of the war repatriated. Forty million dollars invested in the economy would then restore the pre-1967 level. Ward, however, is cautious in reaching conclusions:

> Under the poorest prospects for co-operation in a political settlement which creates a Palestine state on the West Bank, where the state has limited access to Jerusalem's tourism earnings, where trade with Israel is nil, where there is no outlet to the Mediterranean, and where trade even with the East Bank does not achieve its pre-war potential, economic self reliance seems out of the question ... On the other hand, given the most liberal of co-operative agreements concerning the establishment of such a state, including a foreign aid investment plan, resumption of domestic and international trade, commitment of 20 per cent of its GNP to investment, revived remittances and tourism would be rapidly achieved and soon exceed the pre-1967 level. In this environment, economic viability or self reliance for the new entity would soon be

possible.[9]

Ward suggests that the same would hold if Gaza were included, but he wonders whether it is wise to encourage the atomization of industry which would be inherent in an atomized political entity.

Ward's brief study raises more important questions than it answers, and wisely so. The issue of viability is quite complex. His sense of optimism which depends on political co-operation is realistic. Without such co-operation economic viability even under the best of conditions in such a small state would be in doubt. What we question, however, is the implication that the West Bank economy was viable before 1967; if we accept the definition of viability as self-reliance, the West Bank could not have been viable; it was not independent, autonomous or self-reliant.

A more detailed and specialized study of the West Bank has been carried out by Vivian Bull.[10] This study considers three options for the West Bank: federation with Israel, federation with Jordan, or a Palestinian Region in contact with and supported by both Israel and Jordan, and possibly joined by the Gaza Strip, and connected with Jerusalem by special arrangements. Of these, the Region solution is regarded as the most viable economically, while federation with Jordan would be the least viable. The Region solution is considered an embodiment of the benefits of both other solutions, allowing the region an 8 per cent rate of growth which meets the viability criteria of the study. This conclusion is based on the assessment of benefits contained in the Rand study, which may or may not endure if the dependent status of the West Bank relative to Israel is changed. However, a more serious limitation of Bull's analysis is that it does not seriously consider Gaza as part of the Palestine Region, nor does it allow for an inflow of returning refugees which would increase the population by more than the natural rate of 3–3.5 per cent a year. These additional burdens on the resources are bound to become heavier in the case of the Region solution because of the contingent expenditures for government administration, defence and other services inherent in the establishment of an autonomous or independent entity. Hence, Bull's study may have to be qualified even more than it has been by its author to make it relevant.

In contrast to these somewhat sympathetic and optimistic studies, Teddy Preuss has questioned the viability prospects of a Palestinian State in the West Bank and Gaza, and concluded that the economic forecast for such a state is a 'dismal' one. While Preuss' position might seem too partisan, his arguments must be taken seriously. He emphasizes the poverty of natural resources, the cost of maintaining contact between the West Bank and Gaza, the relative isolation from the sea, the current dependence on Israel which would have to be terminated, unless the new Palestine were to continue to be a supplier of cheap labour or a 'labour camp', as he puts it. The alternative to such dependence is a poor Palestine with little prospects of growth and prosperity.

Preuss considers the various suggestions that have been made to support the idea of a state and dismisses them one by one: dependence on employment in other countries such as the oil countries is tenuous and sociologically detrimental to the workers' families; dependence on taxes to support the government is insufficient, given the low expected per capita income in the West Bank and Gaza; dependence on outside aid to sustain the government can only be at the expense of development funds and/or at the expense of political independence; dependence on Israel or Jordan for access to the sea means loss of independence. Thus, in the event of sealing the borders with Israel, 'Palestine will still be an international welfare case'; keeping the borders open would make it a 'labour camp'. Tourism has been considered a major source of foreign exchange; however, since Preuss excludes Jerusalem from the state, he concludes that little tourism can be counted on in the West Bank and Gaza. Preuss then asks with scepticism why the Arab countries should be expected to invest in industry in the Palestine State when they are unable to do so in their own countries. The author summarizes his observations as follows:

It is not easy to discern bright chances for the Palestinian economy. Available data make it plain that continued relations with Israel represents the lesser of the evils, although such a nexus would leave Palestine with the status of a big labour camp exporting its services to richer and more developed countries. The other choice would be dependence upon the charity of strangers, with no independent economic basis. This model would, to be sure, spare Palestine contact with Israel, but at a heavy economic cost which would breed ominous social disorders and would shatter the government set up in Palestine. From the Palestinian point of view, neither alternative is tempting, but that might be the price of independence.[11]

More recent evidence tends to contradict Preuss' observations. It has been estimated that the economic 'burden' on Israel for the years 1968-75 was less than one-half of the one per cent of the GNP a year, 'without taking into account income from oil production in Sinai. . . [and] allowance should also be made for the advantages accruing to this [Israel] country from the conduct of free trade with the areas and from the supply of the types of labor in demand here. It would, therefore, seem that the term, "burden" is inappropriate when discussing economic relations between Israel and the administered areas.'[12]

It would be too easy to dismiss Preuss' conclusions as biased, superficial or undocumented. His logic, however, as far as the viability prospects are concerned, is sound. The economic viability of a State of Palestine must be demonstrated.

In our previous preliminary studies, which we had conducted independently of each other, we approached the problem differently. We estimated the prospective population that might inhabit the State of Palestine and asked: What conditions would be necessary to support this population at a given level of income? Would these conditions be likely to prevail and how? Our preliminary impressions were that these conditions could be realized.

This will be the approach we plan to take in this study. Before we do that, we shall try to establish the logical, political and territorial dimensions of the prospective state within which economic viability will be tested.

Dimensions of the Palestinian State

The number of solutions proposed for a Palestinian state is too large to detail and survey here. For our purposes, only solutions consistent with the idea of two states, one, a Palestinian Arab state, and the other, an Israeli Jewish state, would be relevant. Our approach by definition excludes the extremes of a unified, democratic, secular state as proposed by the PLO, or a Jewish Zionist state extending beyond the boundaries considered legitimate by the United Nations and other international bodies. These extremes are excluded because they deny the full national identity sought by both Arabs and Jews, and because they do not seem feasible or constructive. The solutions we would espouse and recommend must be consistent with the following principles:

1. Both Arabs and Jews have legitimate rights to national identity, self-determination and sovereignty.

2. Territorial war gains cannot be tolerated and must be considered

returnable.

3. A two-state solution within the boundaries of Mandate Palestine is possible.

4. Co-operation, at any level, between the emerging Arab and/or Jewish states or with other parties must be based on the expressed interest of the people of the prospective states. No degree of integration of one state with another or others will be recommended unless it is desired by the people concerned.

The political implications of these basic principles are clear and straightforward, as will be seen below. However, the economic implications may need clarification. Whether in a private market economy or in a planned economy, saving and investment and hence development and growth are closely related to security and incentives. The security of investment and the expectation of success augment the incentives for investment. The above principles are imperative as guarantees of economic security and incentives to invest. They also remove the threat of destruction or evacuation because of war or other military activities. Furthermore, abiding by these principles implies a reduction of the expenditure on the war machinery and a rechannelling of capital into productive investment. Most significant of all is the expectation that these principles would augment the confidence of identity discussed above. The new state and its citizens would be able to enjoy the satisfaction of building their own national home in a secure co-existence with their neighbours.

These principles lead to a two-state solution which gives self-determination or sovereignty to both Arabs and Jews side by side in the land of Palestine. The issue to be resolved is the boundary determination between the two states. Various sets of boundaries may be considered: (1) the 1947 UN Partition Plan; (2) The pre-1967 boundaries; or (3) a new set of boundaries that may be considered rational, secure and technically feasible. Unfortunately not all these boundaries may be considered in detail or as feasible, for different reasons. The 1947 Partition Plan is too far in the past; it suffers from illogical fragmentation of the territory; and it would involve too much dislocation of people to warrant serious consideration.[13] The third proposal, a modified approach to establish rational, secure and technically feasible boundaries would be the most desirable, but only if both parties were inclined toward co-operation and a harmonious negotiated settlement. At present this proposal may seem too idealistic to consider especially in view of the apparent intellectual laziness and lack of creativity in the search for a solution on both sides of the conflict.[14] This leaves us with one practical solution, namely a Palestinian Arab state in the Arab sector of Palestine between 1948 and 1967. This area would include the West Bank, the Gaza strip, and East Jerusalem. These boundaries have been *de facto* legitimized by various UN resolutions, and by the ceasefire agreements between the Arabs and Israel. A possible exception, however, should be noted: the status of Jerusalem has continued to be controversial, since neither the partition of Jerusalem nor its unification under one party or the other has been accepted internationally. The status of Jerusalem, from our perspective, can be resolved in one of two ways: (1) unification, as an international city, or (2) unification, as an autonomous open city under a local civil government.

As an international open city, the authority of Jerusalem would consist of representatives of different nations, presumably under the auspices of the United Nations. Both Arabs and Jews would have equal rights, but neither Israel nor the Palestinian state would claim sovereignty over the city. Freedom of religion and worship would be guaranteed to all.

In contrast, as an autonomous open city, Jerusalem would have a civil authority representing its residents, to be elected by a democratic system of voting. In this case, Jerusalem would be an enclave in which all the freedoms would be guaranteed to the residents, through the good will of the Palestinian state and the state of Israel, and the Kingdom of Jordan. This alternative may also be implemented in the form of two local authorities (sub-municipalities) joined together in an All-Jerusalem municipality. At the same time, both Israel and the state of Palestine may have their capital headquartered in the two parts of Jerusalem, west and east respectively.

Of these two possibilities, the former may be more viable economically since the international authority may contribute to sustain the economy of the city from external sources. However, the autonomous open city model would be more satisfactory politically and socially in as much as it leaves the governance of the city to its people. Both designs, nevertheless, are consistent with the two-state solution and are technically and economically feasible, at least as it seems at this juncture.

In the event that the parties concerned do not agree on the unification of Jerusalem independently of the other governments, the return to the pre-1967 war boundaries would be the only arrangement consistent with the principle of 'no territorial war gains can be tolerated'. In this situation, the restrictions on access to either part of Jerusalem would have to be removed. Both Israel and the Arab state of Palestine may have their capital and administration in Jerusalem and both people may get the satisfaction of free access to their holy places.

The exclusion of Jerusalem from the Palestinian Arab state or from the Israeli Jewish state would necessitate making allowances for the impact on the viability of the given state. The solution we would recommend is to maintain the unity of Jerusalem as an open, neutral, autonomous city, unless the people of Jerusalem elect otherwise. However, since at this juncture the Arabs and the Jews of Jerusalem may prefer to identify with their respective ethnic states, we have considered the partition of the city as in pre-1967 as the premise of our study, pending evidence of more acceptable alternatives.

A virtually identical blue-print of what results from the above determination is contained in the solution proposed by Edward Sheehan, as the most recent in a long list of proposals.[15] A simple map is shown as Figure 3.1 to identify the geographical boundaries which prevailed before the war of 1967 and which would form the boundaries of the State of Palestine, including a highway linking Gaza with the West Bank and a divided Jerusalem.[16] These geographical boundaries contain roughly 2,200 sq. miles, about a fourth of which is cultivated and/or cultivable, unless new sources of water are developed. Population-wise, we should take into consideration various categories of Palestinians who might belong in the state, as in Table 3.1.

Table 3.1: Basic Population Estimates for the Palestine State[a]

Present residents of the West Bank, including East Jerusalem, roughly	682,000
Present residents of the Gaza Strip	430,000
Residents in refugee camps in Jordan, Lebanon, and other places	390,000
Residents of Arab countries who are not citizens, and not in camps	812,000
Potentially returning Palestinians from outside the Middle East	10,000
Israeli Arabs who may elect to reside in an Arab state of Palestine	50,000
Other: Estimated total potential Palestinian population of the new state as of 1975	2,374,000

[a] Figures based on 1975 estimates

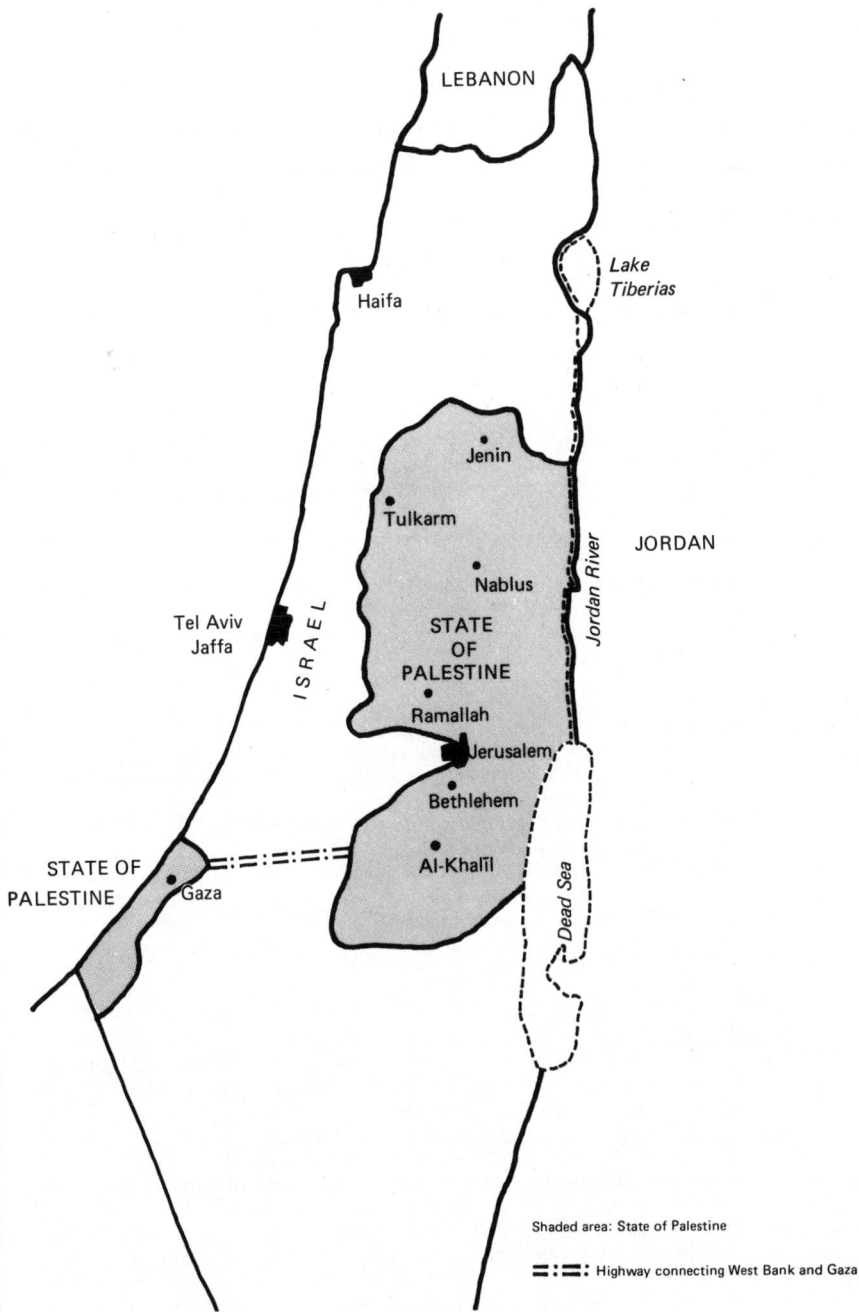

LEBANON

Lake
Tiberias

Haifa

Jenin

Tulkarm

Nablus

JORDAN

Jordan River

Tel Aviv
Jaffa

STATE
OF
PALESTINE

I S R A E L

Ramallah

Jerusalem

Bethlehem

Dead Sea

STATE OF
PALESTINE

Gaza

Al-Khalīl

Shaded area: State of Palestine

Highway connecting West Bank and Gaza

Figure 3.1: Projected Boundaries of the Two-State Solution

Given these area and population estimates, we shall estimate the land, resource and capital requirements to sustain this population at a reasonable level of per capita income, say US$800 per year to start with, what percentage of these requirements exists in the area specified, the amount of deficit, and the means to cover the deficit, such that future growth will be possible to permit the new state to catch up with the highest level attained in the region, as required by the viability criteria discussed above. But first we shall look at the people and the land of the new State of Palestine.

Notes

1. Surveys of the economic conditions of the West Bank and Gaza have been under-estimated, but the only full scale study is concerned with the West Bank alone and does not explore other possible alternatives, as we hope to do; reference here is made to the study by Vivian Bull, to be explored further below.
2. *The Middle East Newsletter,* June-July 1970, vol.IV, nos.4 and 5, pp. 2ff.
3. Haim Ben-Shahar, Eitan Berglas, Yair Mundlak and Ezra Sadan, *Economic Structure and Development Prospects of the West Bank and Gaza Strip,* Santa Monica: Rand, 1971.
4. Ibid., p.9.
5. Arie Bregman, *The Economy of the Administered Areas,* Jerusalem: Bank of Israel-Res. Dept., December 1976, p.28.
6. Arie Bregman, *Economy of Administered Areas,* pp.13-14; similar observations have been reached by Abraham Cohen, *The Economy of the Territories,* Ayn-Hahorish, 1975, (Hebrew).
7. *The West Bank. Social and Economic Structure, 1948-74,* Beirut: PLO Res. Center, Palestinian Book Series #60, 1974 (Arabic).
8. Don Peretz, *A Palestinian Entity,* Washington, D.C.: The Middle East Institute, 1970, p.109; Ward has revised his work but the revised edition retains the same conclusions as the earlier study; R.C. Ward *et al., The Palestine State,* Kennikat Press, 1977.
9. Ibid., p.113.
10. *The West Bank – Is it Viable?,* Lexington Books, 1976.
11. *Davar,* 12 April 1974.
12. Arie Bregman, *Economy of Administered Areas,* p.15.
13. One may argue that the 1947 Partition Plan boundaries are the only legitimate boundaries, given the fact that the UN has not invalidated that plan or recognized other boundaries as legal boundaries. Unfortunately, policy makers tend to minimize the significance of these legalities and to emphasise the pragmatic aspects of the conflict and its possible solutions.
14. Preliminary illustrations of such a solution have been advanced by one of the authors, including maps of the two possible states; *Los Angeles Times,* 1 December 1974.
15. 'A Proposal for a Palestinian State', *New York Times Magazine,* 30 January 1977, pp.8-11.
16. We have not addressed the political issues raised by Sheehan, but have focussed on the economic aspect.

4

THE PEOPLE AND THE LAND

The People

As stated previously, we shall consider the West Bank, including East Jerusalem, and The Gaza Strip as the territory of the new State of Palestine. The population in this area has undergone fundamental changes in the last few years, as Table 4.1 shows.

Table 4.1: Population of the West Bank and Gaza 1968 and 1975

	1968	1975	Growth Rate 1968-73	Growth Rate 1972-5
West Bank	585,500	682,000		2%
Gaza Strip	379,900	430,000		3%
Total[a]	965,400	1,112,000	2%	2.5%

[a] Totals presumably include East Jerusalem, 60,000 in 1968 and 80,000 in 1975.

Source: Estimates by authors from various sources.

The difference in the rates of growth is due to the higher rate of emigration from the West Bank, and the higher natural growth rate in the Gaza Strip. Emigration from the West Bank and its relative absence from Gaza may be explained by local political conditions, access to the East Bank from the West Bank, and the apparent responsiveness of the West Bankers to employment opportunities on the East Bank. The migrants, however, have been of working age and their departure may have left a negative impact on the economy of the West Bank.[1] The pattern of settlement in the West Bank and Gaza may be seen in Table 4.2.

The high percentage of the 0-14 year old group indicates a high degree of dependency. It also signifies the economic burden this age structure will place on the future state in creating employment and generating savings.

The educational background of the West Bank and Gaza has been influenced by various institutional and political conditions. Though occupied by Israel since 1967, the West Bank has continued to benefit

Table 4.2: Pattern of Settlement, 1975

	West Bank	Gaza	Total
Rural	70%	15%	
Urban	30%	85%	
Refugees in camps	73,328	189,997[a]	263,325
Refugees outside camps	125,000	260,000	385,000
Density	110/km^2	1,100/km^2	
Age Structure %			
Age Group			
0-14	48.4	49.6	
15-29	25.4	25.8	
30-44	12.3	12.7	
45-64	9.8	9.1	
65+	4.1	2.8	

a. UNRWA figures July 74-June 75.

from the provisions of Jordanian laws on education. But Israel has also left an impact, while UNRWA has continued to serve as the major provider of education in Gaza. According to the 1967 census figures, a high participation in education has prevailed in both the West Bank and Gaza. Among the adult population, fifteen years of age and above, 70.8 per cent received one to eight years of schooling, while 29 per cent had more than nine years of schooling. Among the refugees, the latter percentage was even higher, reaching 33.4 per cent. The younger age groups had a higher level of education than the older generation, and the male participation in education was higher than female participation. Almost universal education was characteristic of the 15-24 age group, while almost two-thirds of those going to school had over nine years of schooling.

In comparison with both Israel and Jordan, the West Bank and Gaza seem to have a favourable educational basis. The age group 6-11 shows a higher participation in Israel (84.4) than in the West Bank (80.5) but, in the group 15-17 years of age, the percentage in the West Bank is considerably higher than in Israel, 44.6 compared with 22.8 in Israel.

The role of the Jordanian Administration and UNRWA in the progress of education should be emphasised.

Between 1956-57 and 1966-67 (school years) the student population of the country had increased by an average rate of 7.0% while the population growth rate was probably around 3%. The most rapid growth was on the junior high school level (8.7% annually) and the high school level (12.8% annually). On the post-secondary level (university, teachers college, commercial college, nursing school, etc.) the number increased from 800 in 1959-60 to over 5,000 in 1966-7. In addition, over five times that number were studying in universities in foreign countries. By 1966, they had reached enrolment ratios exceeding those of other Arab countries.[2]

In the periods between 1967-8 and 1972-3, the educational level continued to increase by an annual rate of 9,8 per cent. This rate of increase is reflected mainly in the primary and high school levels, although it is observed that the same rate of increase applies to higher education.

The Jordanian Educational Act of 1964 provided free and compulsory education for twelve years. It is interesting to note that this Act still applies to the West Bank, and provides for free and compulsory education up to university level.

While in the West Bank the government played the dominant role in providing education, in the Gaza Strip, UNRWA provided educational services for the large refugee population. The UNRWA schools ensure education for the first nine years but try to encourage an additional three years in secondary schools to prepare the students for higher education. One of the most important achievements in refugee education is the vocational training which graduates a high level of skilled workers, technicians, etc. The supply of technicians, nevertheless, is limited in both the West Bank and Gaza.[3]

The family structure is still traditional in many ways. In 1974 there were about 160,200 families of whom 101,200 families were in the West Bank and 59,000 in the Gaza Strip. The average family in the West Bank consisted of 6.4 persons, with little size difference between the urban and the rural. In the Gaza Strip, the average size of the family was higher and varied: it was 6.8 in the whole area, but it was 6.9 in the urban sector and 6.6 in the refugee camps. Obviously, such large family sizes influenced the general housing conditions, as may be seen in Table 4.3.

The higher percentage of large-sized dwellings in the Gaza Strip in comparison with the West Bank probably resulted from a larger family size. The medium or average density per room was 3.2 in the West Bank and 2.9 in the Gaza Strip. Higher density, however, characterized 32.3

Table 4.3: Housing Distribution per Family

West Bank	% of Families	Gaza Strip	% of Families
1-2 rooms	65.4%	1-2 rooms	55.4%
3 rooms	22.2%	3 rooms	22.6%
4+ rooms	12.4%	4+ rooms	22%

Source: Statistical Abstract — 1976, p.701.

per cent of the homes in the West Bank and 25.5 per cent in the Gaza Strip. Such a high density suggests an urgent need to improve the housing conditions and that housing construction promises to be a major industry in the new state.

The West Bank population includes 18 per cent as refugees. The average density is 110 persons/km². The Gaza Strip has 60 per cent of the population as refugees, and has a density of 1,100 persons/km². The density difference also reflects the ratio between the urban and rural population of the permanently resident population. The urban population of the West Bank is about 30 per cent of the total population (including East Jerusalem), while in Gaza, it is 85 per cent of the total.

The family structure is also influenced by the ratio of males and females. At the end of 1975, there was an equal number of males and females in the total population of 1,112,200. There were 4,000 more males than females in the West Bank, and 4,000 more females than males in the Gaza Strip. However, in the age group under fourteen years, there were more males (51.4 per cent) than females (48.6). In the 15-29 age group the males and females were roughly equal. But there was a considerable difference in the age group over thirty among whom the percentage of females far surpassed the males. This difference may be explained by the outmigration of the head of the family to an Arab country in occupational migration on a temporary basis.

The age distribution determines the potential size of the labour force and, given the technology and institutions, it determines the potential gross national product in the country. Whether the potential is realized or not will depend on the actual level of employment. The labour force participation rate among the population, fourteen years and over, was 34.9 per cent in 1975 in the West Bank compared with 32.1 per cent in Gaza.[4]

One of the factors causing the low participation rate is the very low level of female participation in the labour force. There are no exact figures on women's participation, but it is estimated that between 4,000-5,000 women are employed in the Gaza Strip out of a total of 73,000, or only 6 per cent. However, women employed in local trade and on the farm are not included in the statistics. Nevertheless, the cultural patterns and traditionalism have continued to limit female participation in the labour force.

By comparison, Arab women form 10 per cent of the labour force in Israel, while the percentage of Jewish women reaches 32.5 per cent. Another factor in the overall low participation rate is the emigration of the young, especially from the West Bank, or their pursuit of higher education. Finally, the lack of employment, given the limited economic opportunity in these closed-in areas, must have left an impact on the effective level of participation. This creates a vicious circle; low participation leads to low demand, which in turn breeds low participation in the economic activities of the country.

In 1968, the total number of employed in the local economy was 122,400 and 5,000 in Israel. During that time, 19,000 were unemployed or 13 per cent of the labour force. During the seven years 1968 to 1975 unemployment was liquidated and the additional working population arriving in the labour market received employment, partly in the domestic economy but mostly in Israel. This may explain the decline in the number of employed in the local economy between 1968 and 1975. In 1975, the total number of employed was 205,000, from a total population of 1,112,000, i.e., 18.7 per cent. Of these, 139,000 were employed in the West Bank and the Gaza Strip; and 66,000 were working in Israel. The percentage of employed in the local economy was therefore only 67.8 per cent, while in Israel, 32.2 per cent.

The occupational structure of the local economy, including those employed in Israel may be seen in Table 4.4.

The data for the two years, 1968 and 1975, show that the number of employed in the local economy was relatively stable, having increased only by 7 per cent, in that period, while the population net increase was 17.5 per cent. Another change was a small decrease in the share of employment in agriculture and construction, and an increase of employment in public and private services. It is significant that there was a higher increase of employment in services in the Gaza Strip than in the West Bank. It is apparent, however, that under-employment and disguised unemployment have given a misleading impression of expansion in the service industry. In contrast, the stable employment in agriculture and construction was accompanied by higher productivity

Table 4.4: Employment Structure, 1968 and 1975

	Total Employed in Thousands		Employed WB & GS		Employed WB		Employed Gaza Strip		Employed Israel	
	1968	1975	1968	1975	1968	1975	1968	1975	1968	1975
Agriculture	46	54	45	44	33	32	12	12	1	10
Industry	19	32	18	20	11	14	7	6	1	12
Construction	15	46	13	10	9	8	4	2	2	36
Public services	18	26	18	25	10	16	8	9	–	3
Commerce, transport & private services	37	47	36	40	21	22	15	18	1	5
Total	135	205	130	139	84	92	46	47	5	66
	% of Employed									
	1968	1975	1968	1975	1968	1975	1968	1975	1968	1975
Agriculture	34	26	35	32	39	35	26	26	20	15
Industry	14	16	14	14	13	16	15	13	20	18
Construction	11	22	10	7	11	7	9	4	45	55
Public services	13	13	14	18	12	17	12	19	–	5
Commerce, transport & private services	28	23	27	29	25	24	33	38	20	8
Total	100	100	100	100	100	100	100	100	100	100

Source: A Bregman, *Economy of Administered Areas,* p.25.

and output, enough to cope with the population increase. On the other hand, those who were not absorbed in the local economy seem to have benefited from work in Israel, both in terms of higher earnings and of occupational training.

The Land

The total area of the West Bank and Gaza Strip is about 6,000 sq.km.: of these 5,500 sq.km. are in the West Bank (excluding 200 sq.km. in the area of the Dead Sea) and 367 sq.km. are in the Gaza Strip. There is a considerable difference between the topography of the West Bank and the Gaza Strip. The West Bank is generally mountainous, with a relatively high level of rainfall. Its sub-regions contain the mountains of

Hebron, which are 1,000 meters high, sloping downward toward the narrow strip of the Dead Sea, which is 400 meters below sea level. The Gaza Strip region consists of a narrow plain, which is mostly fertile land and forms a part of the Mediterranean plain, although it contains mainly sand dunes in the south. Land use in the two regions is influenced by the topography and by the historical and political development of each of these two areas.

The West Bank is composed of two main hilly areas (the areas of Jenin and Nablus) and the narrow strip of the Jordan Valley and the Dead Sea Coast. The total length from north to south is 127 km. and the average width is 40 km.

There is an apparent difference between the northern area of Jenin and Nablus and the southern part of the West Bank, containing the Jerusalem and Ramallah regions and Hebron. The northern area is hilly, and is intersected by numerous valleys. It is a mountainous range that slopes down to the north and south and declines sharply towards the Jordan Valley to the east, and less so to the west towards the Mediterranean plain. The Jenin district is lower in elevation reaching 300-400 meters high, while the southern part, containing the Nablus district, is composed of hills about 800 meters high. The Jenin district and the Tulkarm sub-district contain good agricultural land.

The central part of the West Bank, the Jerusalem and Ramallah regions, consist of a moutainous mass. Close by in the south is the Hebron plateau with an average height of 800-1,000 meters; it declines westward to a height of 400-600 meters; it is subdivided by many valleys. Between the mountainous areas of the Nablus, Jerusalem and Hebron districts, and the Jordan Valley and Dead Sea, there is an arid desert area intersected by many valleys and wadis, which leave behind an exceptional morphology and natural beauty within part of the land. In spite of the high temperature, the Jordan Valley and the Dead Sea coast, in addition to their natural beauty, have high potential for agricultural land use.

These topographical conditions have influenced the present land use and the location of human settlements. They no doubt will also influence the prospects and patterns of future development. The patterns of present land use are partly due to the technological conditions which have prevailed during a long period of history when agriculture was the dominant factor in the national economy. Another factor which has influenced the land use patterns of the West Bank is the distance from the Mediterranean, which has rendered the area dependent on the coastal towns for marketing and trade purposes.

Table 4.5 shows the distribution of space and population by district. These data show that a high percentage of the rural population is concentrated in the Jenin and Tulkarm districts while the percentage of urban population is very low. The districts of Ramallah and Bethlehem show about 75 per cent of the population as rural, while the lowest percentage of rural population is seen in the Nablus district.

The population distribution corresponds to the distribution of the agricultural land the space used by the 400 villages and 11 towns. The evaluation of land used for agriculture, human settlement and other uses is important for estimating the unused land or land potential available for future development.

Table 4.5: Distribution of Population According to the Districts and Corresponding Areas

District	Space in sq.km.	Population					
		Total	Urban	%	Rural	%	
Jenin	572	91,045	13,365	15	77,680	85	
Tulkarm	332	100,631	15,275	15	85,356	85	
Nablus	1,587	108,692	44,223	41	64,469	59	
Ramallah	770	101,634	25,532	25	76,102	75	
Jerusalem	284	60,000	60,000	100			
Bethelem	565	67,482	16,313	24	51,169	76	
Jordan Valley	338	10,795			10,795	100	
Hebron	1,056	118,358	38,309	32	80,049	68	
Total	5,504	598,637	153,017	26	445,620	74	

Source: Elisha Efrat, *Judea and Samaria: Guidlines for Regional and Physical Planning,* Ministry of the Interior, Planning Dept., Jerusalem, 1970, p.9. *Statistical Abstracts,* 1968.

The total space of cultivated agricultural land is approximately 2,000 sq.km. (200,000 hectares). The quality of the land varies according to topographical differences and soil structure. Four categories of agricultural land may be distinguished: (1) deep soil with a slope of up to 4 per cent; (2) deep soil with a slope of between 4 and 15 per cent; (3) soil limited in depth with a high percentage of stones and with a slope above 15 per cent; (4) mountain areas with a thin layer of soil for extensive cultivation of olive trees and vineyards.

About 50 per cent of the total cultivated area belong to the last category, while only 1,000 km^2 (100,000 hectares) might be considered as good agricultural land in the present state of technology.

Table 4.6 shows the geographical distribution of the agricultural land according to the different soil categories. The categories of the regions do not correspond to the governmental geographical classification of districts.

Table 4.6: Cultivated Area According to Categories and Regions (in sq.km)

Region	Total	A	B	Categories C	A-C	D
North	600	190	30	180	400	200
Central	900	120	60	120	300	600
South	300	50	60	120	230	70
Jordan Valley	200	140			140	60
Total	2,000	599	150	420	1,070	930

Source: Estimated by authors

These data demonstrate that most of the agricultural cultivated land of the first three categories is concentrated in the northern and central regions. A high percentage of the 'A' category is located in the northern region which corresponds to the administrative districts of Jenin, Tulkarm and parts of Nablus The large area of the Hebron district has only 50 km^2 of the 'A' category, or only 17 per cent of the total cultivated land in the district. The central regions including Ramallah, Jerusalem and parts of the Nablus district contain a significant part of category 'A' but, at the same time, two thirds of the cultivated area belong to category 'D'. The region of the East Valley contains the smallest percentage of cultivated land, however: almost this entire area belongs to category 'A'.

A more detailed analysis of the techniques and uses of agricultural land by crop and level of output may show that some part of the land within category 'D' may be used for purposes other than present usage. Such reallocation may increase the output of the remaining parts of the land by increasing the efficiency of cultivation.

There are some difficulties in estimating the land space presently utilized for urban and rural settlements. Table 4.7 shows the land space and population within municipal borders. These figures do not necessarily show the land in urban use. The following figures show land classified as within municipal administration.

Table 4.7: Urban Population and Space in Municipal Administered Lands According to the 1967 Population Census

	Population	Space Km²
Hebron	38,091	74.37
Bethlehem	14,439	7.50
Beit Jala	6,041	9.37
Ramallah	12,031	19.70
Ve Bira	9,568	22.50
Jericho	5,200	37.80
Jenin	8,346	19.37
Tulkarm	10,157	6.25
Kalkilya	8,922	7.80
Nablus	41,537	—
Total	155,235	

Source: Elisha Efrat, *Judea and Samaria,* p.9; Abshalom Schmueli, *Hebron, Patterns of a Mountain Town,* Tel-Aviv; Gama Publishing House, 1970.

Recent research on the development of Hebron shows that the town of 40,000 inhabitants and 70 km² of land within the municipal administration, has no more than 6 km² in urban use. The 63 km² in agricultural use, although under the municipal administration, are not available for urban growth. The *Wakf* (religious trust) which owns the land does not allow any changes in its use. The 6 km² in urban use are unequally distributed among the inhabitants of the town. Land use patterns and population distribution of this town are similar to most Middle Eastern cities — the central area together with the old market and the old commercial centre are densely populated, while the neighbourhoods erected during a later historical period show a much lower density.

The average allocation of land for all urban purposes in Hebron is 150 m²/person. This average may serve as a basis for estimating land space use in other towns in the West Bank, even though there are differences in space distribution and patterns of development among the different towns.

The total urban population in 1968 was 166,000 in all 11 towns excluding the 20,000 inhabitants living in refugee camps within the municipal zones, and the 66,000 inhabitants in East Jerusalem. According to the average of 150 m^2/person, the space occupied by the urban population 175,000 was 26 sq. km. Allowing for the unused or vacant land within the urban settled areas, it is reasonable to increase this figure by 50 per cent. The land in urban use may therefore be estimated at about 40 sq.km.

It is equally difficult to estimate the land used by the 400 rural settlements: 7 per cent of the villages are large settlements of over 3,000 inhabitants and contain 25 per cent of the rural population. The density in these villages is higher than in the smaller villages. Land allocation for rural settlements may be estimated at 250 sq. metres per person or 1,500 m^2 for a family of six persons. The land allocation for rural settlements does not include the land in garden use located near the house of the farmer.

On the basis of the established norm of 250 m^2/person, the land used for a rural settlement of 400,000 persons may be estimated at 100 sq.km. Thus, the total land allocation for human settlement is approximately 140 km^2 (40 km^2 in towns and 100 km^2 in villages).

In order to estimate the available land for future development, it is necessary to evaluate the land in these other uses: (1) roads; (2) mineral and other natural resources; (3) nature preservation; (4) historical sites; (5) land available neither for agriculture, nor for human settlement due to soil conditions (desert, rocks, etc.).

According to surveys in different countries, the land designated for roads and railways is 50 per cent of the land used for human settlements. According to this estimation, we may consider 70 km^2 of the land to be used for roads. Probably this figure is higher than the actual space used for roads, since a large part of the road network is located in rural regions connecting the villages with the towns, which are included in the classification of agricultural land use.

So far we have no evidence of a substantial quantity of soil strata in the mountain area suitable for a building material industry. The mountains in the areas of Hebron, Bethelehem, Jerusalem, Ramallah and Nablus are rich in different kinds of calcareous rock and other suitable building material. There is an abundance of quarries in these areas and they are exploited in a restricted way solely for construction purposes.

The most important minerals for use in future development are the rich resources of potassium and other chemical elements existing in the Dead Sea area. It should be noted that the north end of the Dead Sea was the starting point for development of the Dead Sea resources. The south end of the Dead Sea Development Plan was only constructed in later stages of development, and eventually became the centre of the developed petrochemical industry of Israel. Recently, Jordan has invested large sums of capital to develop the East Dead Sea resources. Through appropriate investment there are possibilities of renewing the operations in the north of the Dead Sea.

Different areas of the eastern side of the hills which slope sharply to the Jordan Valley and the Dead Sea, together with some rich vegetation areas in the Jordan Valley, and some water sources in the Dead Sea area are considered worthy of natural preservation. The best known among these areas are: Ein Feshka and its surroundings, Kirmizan near Beit Jala, the upper flow of the Jordan River and its peculiar vegetation, and the numerous wadis (Wadi Kelt) which cross the desert slopes of the northern district. This land area may be estimated, according to present knowledge, to be about 50 sq. km.

Numerous areas among the mountainous regions of Nablus and Jerusalem and its environs are significant for religious and tourism purposes. These include: Sebastia, Tel Balata, Sartala, Quman, Mount Hordua, Nebi Mussa, Mar Scibba, the monasteries of Wadi Kelt, etc. The entire area of the historical sites and their road connections comprises about 50 sq.km.

Finally, there is land which due to natural soil conditions cannot be used in future development. The desert area of the southern and central districts, consisting of a series of peaks and terraces sharply sloping to the Dead Sea and to the Jordan Valley, covers an area of about 800 sq.km. not suitable for human settlement. In addition, 200 sq.km. of wadis and rocks located in the mountainous areas of the Nablus, Jerusalem and Hebron districts are inappropriate for human settlement purposes. The total of these areas may be estimated to be about 1,000 sq.km. although some parts of these areas may be used for nature preservation, national recreation purposes and tourism. These estimates are shown in Table 4.8.

The Gaza Strip is located alongside 42 km. of Mediterranean Coast with a width of 8 to 10 km. The total space is 360 sq.km. This is a plain area mostly for agricultural use: 200 km^2 is cultivated, 45 per cent of which is irrigated.

The cultivated area is settled by 31,000 permanent rural population and 40,000 refugees living in rural areas. Eighty per cent of the total population lives in urban districts of the towns of Gaza and Khan Yunis. Gaza contains about 100,000 inhabitants as permanent residents and

Table 4.8: Land Use Distribution Estimates in the West Bank

Agricultural use:		2,000 sq.km.
Human settlements:		
Urban	40 km^2	
Rural	100 km^2	140 km^2
Other uses:		
Roads	70	
Natural resources	30	
Nature preservation	50	
Historical sites	50	200 km^2
Unusable (roads, desert)		1,000 km^2
Total		3,340 km^2
Land reserve		2,200
	Total	5,540 km^2

63,000 refugees living in camps on the outskirts of the towns. Khan Yunis is populated by 50,000 permanent population and 104,000 refugees on the outskirts.

On the basis of land used by urban and rural settlements in the West Bank, the following data seem to be good approximations of the present norms: 150 m^2/person for urban population; 250 m^2/person for rural settlements; 70 m^2/person in refugee camps. Accordingly, an estimated 40 km^2 of the urban and rural space are used for human settlement by the permanent and refugee population in the Gaza Strip. The roads occupy about 20 km^2. The total space used for agriculture and human settlement is 260 km^2; 200 km^2 for agriculture and 60 for human settlement. According to our rough estimation, there are 100 km^2 more in the Gaza Strip that are also usable, presently in the form of sand dunes. This raises the total potential usable land in the new state to 2,300 km^2.

Notes

1. This theory, however, may be challenged, given the continued lag of the economy of the Gaza Strip, even though no such exit of manpower has taken place.
2. E. Kanovsky, *The Economic Impact of the Six Day War*, New York: Praeger, 1970, p.96.
3. For more details, United Nations, General Assembly, *Report of the Commissioner-General of the United Nations Relief and Works Agency for Palestine Refugees in the Near East*, A/10013, 1975.
4. A. Bregman, *Economy of Administered Areas*, Jerusalem: Bank of Israel, 1976, p.24.

5 THE ECONOMY

It is rather difficult and probably misleading to study the economy of the West Bank and Gaza as an independent unit and base forecasts on the study since no such autonomy has existed before. The impact of Jordan and Israel has been overwhelming. Nevertheless, a look at the various sectors should give an idea of what exists, and of what has been happening in that economy.

Agriculture

Agriculture is the main sector in both the West Bank and Gaza. In 1975 agriculture employed 32 per cent of the labour force, with 35 per cent in the West Bank and 26 per cent in Gaza. It contributed about 30 per cent of the GNP in the West Bank and 29 per cent in Gaza. The role of agriculture, however, especially in the West Bank, goes beyond its contribution to the GNP and employment since about two thirds of the people live in rural settlements. A large part of those employed in the service sector live in the countryside. They earn their living by serving agriculture by marketing tools and inputs to the farmers, and agricultural produce to the urban centres.

The development of agriculture during the period between 1968-75 has been influenced by the structural differences between the two areas as well as by the increases of output and income due to more efficient use of the agricultural land.

The essential difference between the West Bank and Gaza Strip is demonstrated in the gap between the size of the cultivated area in the two regions and the disproportionately smaller difference in the value of agricultural produce. In 1975, the total cultivated area in the West Bank was 2,022 sq.km. while in the Gaza Strip it was only 210 sq.km. The value of the agricultural produce in the West Bank was IL 592 million (in 1974 prices) — while in the Gaza Strip, it was IL 257 million. The output per sq.km. in the West Bank was IL 296,000, while in the Gaza Strip it was IL 1,223,000 or, four times higher in the Gaza Strip than in the West Bank.

The main reason for such a difference appears to be irrigation: in 1973, the total irrigated area in the West Bank was 81 sq.km., or 4 per cent of the total cultivated area, while in the Gaza Strip it was 95 sq. km. or 45 per cent of the total cultivated area. The main crops and their relative significance are shown in Table 5.1.

Table 5.1: Land Use Distribution by Crop: West Bank (thousands of dunams[a])

	1968	1973
Total Cultivated Area	2,045	2,017
Field Crops	890	827
wheat	465	430
barley	231	150
vetch	48	30
sesame	18	40
lentils	61	60
chick peas	32.5	80
tobacco	4.5	6
others	30	31
Orchards	680	765
olives	514	580
grapes	54	62
citrus fruits	24	24
figs	38	15
almonds	34	61
others	17	23
Melons	43	10
Vegetables	70	80
In preparation	362	335
Included in the above:		
Irrigated area	57	81
vegetables	31	54
citrus fruits	24	24.5
bananas	2	2.5

a. A dunam is approximately one-quarter of an acre.

Source: Van Arkadie, *Benefits and Burdens*, p.129.

According to these figures, the field crops occupy about 40 per cent of the total cultivated areas; fruits about 38 per cent; vegetables only 4 per cent. Most of the vegetables are grown on the irrigated land. One of the most striking features is the share of olives, which occupy about 29 per cent of the total cultivated land. A different picture of the agriculture is seen in the Gaza Strip. The citrus orchards are the most important speciality, and occupy one third of the total. About 22 per cent of the cultivated area is devoted to such crops as almonds, grapes and olives. Vegetables comprise a large and important part of the economy, similar to that of citrus. In terms of value product, Table 5.2 illustrates the essential difference between the two areas. These data show that field crops comprise 11 per cent of the total value agricultural output in the West Bank, though they occupy 40 per cent of the total cultivated area. On the other hand, 4 per cent of the cultivated land in vegetables contributes 20 per cent of the output.

Table 5.2: Crop Distribution of the Agricultural Produce in 1975 (per cent value)

	West Bank	Gaza Strip
Field crops	11	1
Vegetables	20	10
Fruits	31	61
(Olives or citrus)	(7) Citrus	(51)
Melon	—	2
Livestock	37	26
Their produce		
Forestry and others	1	—
Total	100	100

Source: Bregman, *Economy of Administered Areas,* p.42.

The contrast between irrigated and non-irrigated areas shows how vulnerable dry farming in this area is. The variation in the value of real output will illustrate, as in Table 5.3.

The most striking fluctuation begins in 1972 which shows a 37 per cent increase in value of agricultural product over 1971; 1973 shows a 16 per cent decline, while 1974 has an increase of 64 per cent over 1973, followed by a 34 per cent decline in 1975.

In the Gaza Strip, the average yearly change in the value of output is the same as in the West Bank, or about 10 per cent. But there is no such sharp variation between the different years. The period 1969-1972 shows a 10.14 per cent variation, while 1973-5 shows variations of 3.7

Table 5.3: Change in Real Value of Agricultural Output and Real Income of Farmers (per cent change)

Year Relative to Previous Years	West Bank		Gaza	
	Value of Agr. Product	Income	Value of Agr. Product	Income
1968/69	26	31	14	9
1969/70	−10	−14	13	30
1970/71	14	38	14	35
1971/72	37	25	10	16
1972/73	−16	− 7	7	− 1
1973/74	64	76	6	− 5
1974/75	−34	−39	3	33
Average annual change 1968/75		14		15

Source: extracted from Bregman, *Economy of Administered Areas*, p.41.

per cent. In general the Gaza Strip has experienced a relatively steady increase in the value of agricultural product, while the West Bank has witnessed sharp increases or declines from one year to the other. Obviously the differences in rainfall strongly influence the fluctuation in field crop output, especially olives which constitute 29 per cent of the total cultivated land in the West Bank.

The growth of agricultural output and income may be explained in part as a function of change in the use of inputs. The number of tractors increased from 130 to 1,220 during the period 1968-75; in contrast, the number of horses and donkeys used in agriculture decreased from 5,400 to 2,600 in the same period. The value of fertilizers used increased from IL 1.8 million in 1968 to IL 6.5 million in 1973. The value of pesticides and similar inputs increased from IL 0.7 million to IL 6.3 million in that same period. Similarly, there was a rise in the value of feed from IL 10.5 million to IL 33.3 million.

Another factor in the rise of the value of output has been a more efficient system of marketing of both agricultural inputs and outputs. As a result, there has been a rise in the export of some crops, not only to the neighbouring countries, but also to the European market, which is a new market for the West Bank and Gaza, except for citrus.

However, the change has also been evident in land yield, as it was in labour productivity. The wheat harvest, for example, rose from an average of 1,000 kg. per hectare to 1,500 kg. by 1973; the barley yield rose by the same ratio; the yield of onions rose from a range of 1,000-5,000 kg. per hectare to a range of 12,000-17,000 kg. per hectare by 1973. A similar rise in yield has been observed in other crops such as sesame, tobacco, melons and some vegetables as well as in the citrus yield, although at a lower rate.

In contrast, there has been a very small change in the livestock sector which constitutes a third of the value of agricultural product. The technological basis of this sector did not change during the recent years. Feeding has continued to depend on natural pasture. There has been a very limited use of feed mixtures and veterinary services. As a result, the shortage of meat supply and of milk and milk products has continued to characterize the economy.

Finally, it should be noted that the progress achieved in the past has been the result of work carried out by a small number of people, consisting of about 500 local agricultural experts and 19 Israelis. This staff has succeeded in achieving major results, in spite of the limited means at their disposal. Few farmers have access to the new techniques, and demonstration plots have been sparsely scattered in the area. There are, however, a few more obstacles in the way of advanced agriculture, which will be mentioned briefly.

First, due to tradition, consumption habits and the current methods of production, the farmers continue to grow corn for animal feed instead of sorghum, even though sorghum has been shown to give higher yields and income. Second, the land is still fragmented in small plots, which makes mechanization difficult, even for crops such as sugar cane and beets which require large-scale farming. Third, severe water scarcity in the West Bank has continued to be the major limiting factor in agriculture. Finally, while marketing has improved, there is yet no organized system of marketing so as to co-ordinate production with the market conditions.

Given these handicaps, it is reasonable to suggest that removal of these obstacles would revolutionize agricultural production in the area.

Industry

The role of the industrial sector in the economic structure of the West Bank and Gaza has been small and limited to meet domestic demand. Both the West Bank and Gaza have depended heavily on the dominant administration for industrial supplies.

As late as 1965, manufacturing in the Jordanian economy contributed only 10 per cent of the GNP. However, the share of industry in the

West Bank was even lower than in the East Bank. It contributed only 7 per cent of the total GNP and employed 15,000 workers. After the 1967 war, industrial activity was considerably reduced. The number of people employed in industry declined to 9,000 in 1968, rose to 12,000 in 1969, and to 14,000 in 1975, or to about 10 per cent of the labour force employed domestically. Food industries employed 25 per cent of the industrial labour force while textiles and clothing accounted for 14.6 per cent of the employment.

In the Gaza Strip, industry played an even smaller role in the economy. According to the 1967 census, 6,100 people stated that they had worked in industry before the war. This figure fell to 2,700 after the war, increased to 7,000 in 1968, but declined again to about 6,000 in 1975. The main branch was clothing and textiles, but 39 per cent of those working in industry produced mostly for Israeli enterprises. Manufacturing of food represented 10.6 per cent of the industrial labour force.[1]

There has been no structural change in industry under the Israeli administration. The impact of reduced consumer demand on the domestic market and the lack of investment incentives to enlarge the industrial sector are in part responsible for this relative stagnation. Additional factors include the conditions under which the weak local industry must compete with the developed Israeli industry, a lack of developed financial institutions, and the political uncertainty surrounding the future of the area.

Generally speaking, one of the major factors of industrial development in less developed countries is the government sector. But such government initiative has never existed in the West Bank or Gaza under any recent administrations. Schemes were prepared by the Israeli authorities for the encouragement of industrial development. But rather than promote industrial development in the West Bank and Gaza, the authorities chose the alternative of importing manpower from these areas in order to strengthen further the Israeli economy. Thus they guaranteed the supply of relatively cheap labour for the industrial enterprises, construction work and agriculture in Israel. Industrial production, nevertheless, increased in both the West Bank and Gaza by about 80 per cent between 1968 and 1975, while the GNP increased by 120 per cent. From 1968 to 1975, the share of industrial output declined from 7.9 per cent to 6.4 per cent in the West Bank, but rose from 3.7 per cent to 5.7 per cent in Gaza. The increase in Gaza was mainly in local workshops which produced on contract for Israeli enterprises.

The difference between the relatively high increase of industrial output and the small change in the number of employed in industry suggests a significant increase in productivity and a more efficient use of manpower. This may have been due to a reduction of the under-employed and disguisedly unemployed, in response to the increasing demand for labour in Israel. It may also have resulted from increased technical exposure to and competition with the more advanced Israeli industry.

Construction

The building activity in the West Bank and Gaza has undergone many changes during the past few years. The number of people employed in construction in the West Bank during the Jordanian administration reached 25,000, but it decreased to 10,000 by 1968. By 1975, only 8,000 people worked in construction, although 36,000 others went to work in the building industry in Israel.

While the number of employed declined, the output of the construction works increased by 32 per cent between 1968 and 1975 in the West Bank and by 30 per cent in the Gaza Strip. However, the contribution of construction to the GNP in the West Bank increased from 4.2 per cent to 10 per cent during the years 1968 to 1975, and from 4.8 per cent to 10.4 per cent in the Gaza Strip. It is interesting that expansion in the building industry was concentrated in housing. It may be suggested that the progress in the housing industry was the result of an increase in the GNP per capita and the investment of savings in housing rather than in production branches of the economy.

It may be striking that the high rate of increase in building activities was not accompanied by an increase in the number of people employed in the industry, and sometimes was even accompanied by a decline. The reason lies in the higher productivity of labour and in the fact that a substantial part of the construction work was accomplished by unpaid family labour. Nevertheless, the residential density still characteristic of the area suggests that the construction industry has not expanded enough to meet the demand of the present population.

Services

The services sector included a wide range of economic activities such as transportation, commerce, tourism, public services and various kinds of private services. The inter-relationship between the different parts of this sector depends on the economic structure of the economy and the level of ecomomic development. In general, industry generates more demand for services than does agriculture, which may explain the relatively high percentage of people productively employed in the service industry in the developed countries, compared with the developing countries. However, in some developing countries, the role of the service industry may be high because of the lack of alternative employment and the ability of that sector to disguise unemployment.

Prior to 1967 the share of services in employment was declining slightly over the years. In 1965, the services sector employed 33.3 per cent of the labour force in the West Bank while its contribution to the GNP was close to 62.5 per cent. The relatively high contribution to the GNP corresponds to the expanded tourist industry, to commerce, and the higher labour productivity in that sector relative to other sectors.

In the Gaza Strip, the services sector has played a major role in employment, with 55.7 per cent of the labour force engaged in the services. The corresponding contribution to the GNP was 55.1 per cent. Apparently there are more public services in Gaza than in the West Bank, including the UNRWA services in the refugee camps which may have lower productivity than tourism.

The service industry has continued to play a major role. In 1975, the service industry employed 41 per cent of the labour force compared to 37 per cent in 1968 in the West Bank. In the Gaza Strip, employment increased from 50 per cent to 57 per cent in that period. However, there has been a decline in the services' contribution to the GNP between 1968 and 1975. In Gaza it declined from 67.7 per cent to 35.5 per cent in that period. Apparently there has been a substantial increase in disguised unemployment and under-employment in the service industry, as has been common in the underdeveloped and developing countries.

The Aggregate Economy

The economic activities of the West Bank and Gaza as at the end of 1975 may be summarized by Table 5.4.

It is easy to see the difference between the per capita GNP in the West Bank and Gaza. The high number of refugees in the Gaza Strip compared with the West Bank probably contributes to the difference. The higher share of income from abroad coming to Gaza helps to mitigate that difference but not to eliminate it.

In general public investment has played a negligible role in both the West Bank and Gaza, compared with the private investment. Investments in the public sector reached IL 82 million in the West Bank in 1975 and IL 77 million in Gaza. Private investment amounted to

IL 382 millions and IL 128 millions in the two areas respectively.

Table 5.4: National Accounts of the West Bank and Gaza Strip in 1975 (million Israeli Pounds[a])

	West Bank	Gaza Strip
GDP	2,465	2,351
GNP in market prices	3,242	1,465
GNP at factor prices	3,133	1,428
GNP per capita (IL)	4,764	3,415
Private consumption expenditure	2,780	1,282
Public consumption expenditure	294	168
Gross domestic capital formation	514	566
Export of goods & services	788	566
Subsidies to export	23	41
Total resources	4,399	2,351
Less import	1,877	1,302
Less taxes on import	57	9
GDP	2,465	1,040
Factor payments from abroad	801	430
Less factor payments abroad	24	14
GNP at market	3,242	1,456
Less indirect taxes on domestic production	109	28
GNP at factor prices	3,133	1,428
GNP per capita (IL)	4,764	3,415
Private consumption per capita (IL)	4,085	3,007

a. The effective rate of exchange $1 = IL 7.16 in 1975, though the official rate was 6.3275 according to the Bank of Israel Annual Reports.

Source: Statistical Abstract of Israel 1976, pp.690, 691.

As may be expected, the economy of the West Bank and Gaza suffers from a deficit in the balance of trade, with imports exceeding exports by a ratio of two to one. The relationship between imports and exports is similar to what it is in Israel, only more serious. The trade deficit in Israel amounted to 28 per cent while in the West Bank the deficit is close to 40 per cent and to 70 per cent in Gaza. But there is also a difference in the way the deficit is financed. The West Bank and Gaza cover the deficit through income and aid from abroad, while most of the Israeli deficit is financed through long-term borrowing.

The overall picture of the economy of the West Bank and Gaza Strip is one of a relatively underdeveloped economy. It is dependent to a large degree on the economies of Jordan and Israel, not only to finance the trade deficit, but also for the services of financial and trade institutions. The monetary system is in fact a confused one: two currencies, Jordanian and Israeli, are legal tender. Branches of Jordanian banks exist but do not function. Israeli banks have branches but are not trusted nor are they committed to the servicing of the area as they do within Israel. Local banking and financial institutions have not evolved, for various reasons, including the insecurity of being politically and economically dependent. As a result, the channelling of resources into productive investment has not been smooth or easy, nor has it been conducive to modernization and productive investment in the economy. There is evidence of hoarding in both the West Bank and Gaza by anywhere between 5 and 15 per cent of the private income. Thus, while underdevelopment and relative backwardness of the economy are evident, a high potential for saving and investment exists. Whether the independent state conditions will release these suppressed capabilities and channel them into economic viability remains to be seen.

Note

1. Vivian Bull, *The West Bank – Is It Viable?*, pp.95-6; A. Bregman, *Economy of Administered Areas*, p.28.

6 THE VIABILITY PROSPECTS I – SECTOR CONTRIBUTIONS

Economic forecasting usually depends on past experience and general trends in the economy. With allowance for the changing conditions, estimates of the rates of growth of income and employment, and of price change are made and then used to forecast the future trends. Our study cannot apply the same method because of the total discontinuity embodied in the creation of the new state of Palestine. Geographical boundaries will be redrawn; the government structure and political institutions will be altered; and the population size and composition will be radically changed. Hence, no past trends can be utilized to predict future trends. Instead, we shall take certain parameters as given, such as the population, the land and the goals of the Palestinian State and proceed to assess the degree to which those goals may be realized. We shall look at the various economic sectors individually and in combination.

The People and the Objectives

We have surveyed briefly the demographic and economic features of the people residing in the West Bank and Gaza, as of 1975. These, however, will constitute no more than 50 per cent of the prospective population of the new state during the first few years of its existence. Who are the other 50 per cent? Where do they come from, and what are their goals and capabilities? How ready are they to play the necessary role to create a viable economy?

As noted above, we can only guess the answers to these questions. The returning people will be Palestinians living in refugee camps in other countries of the Middle East, Palestinians in the Arab countries but not in camps, nor are they citizens, and Palestinians residing outside the Middle East who wish to return upon establishment of the state. In addition, we assume that a certain number of Israeli Arabs from Galilee, Haifa, Nazareth and other places in Israel will want to come and live in the new state. Estimates of these groups have been made above, as far as their numbers are concerned. But what are their characteristics?

The Palestinians of the refugee camps are the majority of Palestinians outside the West Bank. They are concentrated in Lebanon, Jordan and Gaza. To a large extent, they have depended on UNRWA for education, health and a modest support of their daily living. They are the most anxious to establish a state, the most affected by displacement, and probably the most determined to make a success of the new state. While detailed profiles are not available, we have strong indications that the Camp Palestinians have in the last decade rechannelled their energies toward resolution of the conflict and toward qualifying themselves to be of service and advance the new state. As a result, school attendance has been high at all levels; advantage has been taken of the vocational training centres available to them. And many of them have maintained their relationship with the camps, even after they have become professionals and in great demand in various economies around the world. Many of them have kept their contact with the camps as leaders, whether in politics, social services or military activity. This professional group has also been the bridge between the Camp Palestinians and the outside world since they are the most mobile group of Palestinians in the Middle East region.

In contrast to the Camp Palestinians there are a fairly large group who have lived in urban centres in Lebanon, Syria, Jordan and Egypt, but who have not become citizens. These people have probably been the most isolated since they are neither close to other organized Palestinian groups, as in the camps, nor have they been assimilated like those who have acquired citizenship. They no doubt would consider returning to Palestine, but it is doubtful that they would rush into a living situation that may be less certain than the situation they are in presently. However, to the extent that they have not acquired citizenship, they no doubt are good candidates for eventual return. It may be noted that a large percentage of these people have been able to earn a living on their own. Most of them are technicians, small businessmen and craftsmen. Their return would be an asset to the new State of Palestine.

An important, though small, group are those who have acquired citizenship, assimilated into the society in which they happen to be, and have achieved professionalism, wealth and positions of leadership. These people keep in touch with other Palestinians through intellectual and financial contributions to the cause of Palestine. Among the members of this group are large-scale contractors, engineers, bankers, medical people and social scientists. Whether these people would return to Palestine is uncertain. However, the impression one gets from exploring their intentions is that they will continue to lend their support and that many will return on an exploratory basis before making a final decision.

Finally, there are Palestinian Arabs who are citizens of Israel. These are the least organized, most demoralized and ambivalent toward the Arab-Israeli conflict. Whether many of them will choose to move to the new state is difficult to tell. The early experiences of the new state will determine how many will be attracted over. The level of vocational and professional training among the Israeli Arabs is lower than among the other Palestinian groups. In fact most of the professionals among the Israeli Arabs have either left Israel or are waiting for a chance to leave – hopefully to the western world. The contribution of the Israeli Arabs to the new state may come in two ways: by displaying a non-traditional attitude toward work and economic enterprise, and by serving as a bridge for pacification between the State of Palestine and Israel.

According to a survey of Palestinian manpower in the early 1970s, it appears that per capita university education among the Palestinians has been high, relative to all other countries in the Middle East. Absolute numbers are not available, but an estimate of 50,000 university graduates by 1969 has been advanced. Assuming the number has continued to expand by 10 per cent a year, the number in 1977 will have passed 100,000.[1]

It is more significant, however, to observe the specialization of the university graduates. While a large number of undergraduates concentrate on the humanities, the graduates tend to concentrate on medicine, engineering and science. The more advanced the degree, the higher the frequency has been of entering science and engineering, as Table 6.1 shows. Medicine is by far the most popular field among advanced degree holders, reaching almost 60 per cent of the total.[2]

Table 6.1: Distribution of Palestinian University Graduates as of 1969 (per cent of total)

	Humanities	Science
	%	%
BA	70	30
MA	54	46
Ph.D.	16	84
Other	6.6	93.4
Total	61	39

Source: N. Shaath, 'High Level Palestinian Manpower,' *J. of Palestine Studies*, vol. I, no.2 (Winter 1972), p.85.

A specific group that should be mentioned are the young Palestinians, males and females, who were in the armed forces of the Palestinians. Whether in Fath or in Saiqua or in the Palestine Liberation Army, these people have been disciplined and have acquired a variety of training specialties. They will probably continue to serve in the armed forces of the new state, whether for local law and order, or for a small standing army, or for a vanguard in the resettlement activities. The energies, discipline and training of these young returnees should be invaluable in making the new state a viable entity. To channel their abilities into constructive activities will be a challenge but the returns no doubt will be high.

Having identified the prospective population, let us specify the economic objectives that are consistent with viability. The objectives consist of productive employment, an acceptable level of income per capita, and the potential for saving and investment or a rate of growth that is comparable to the growth rates of other countries in the region.

Given a population of 2,374,000, a participation rate of 23 per cent, we will have a labour force of 546,000. Assuming 4 per cent structural unemployment at the full employment level, the economy must generate 524,160 jobs in all sectors combined. If we agree that a per capita income of $800 is reasonable within the first five years, the economy must at the full employment level produce about $1.9 billion a year. To maintain a favourable rate of growth, the economy must generate savings and investment to offset the population growth and expand the output by no less than 10 per cent a year. The alternative would be to control the population growth as a way of making it easier to meet the saving and investment targets from domestic sources. We shall continue the analysis on the basis of these estimates for simplification. However, a true picture must reflect the population growth and the rate at which the returning Palestinians will be repatriated. We shall make these modifications in the next chapter. In the meantime, we shall estimate the contribution of each sector to employment and the total output in the economy.

The Contribution of Agriculture

The increase of agricultural output during the period 1968-75 at 8-10 per cent a year due to the introduction of more efficient methods of cultivation and land use suggests that if similar measures could be extended to all cultivated areas, then it should be possible to double the agricultural output in a period of 7 to 8 years. It will also be possible to raise the income per capita from agriculture substantially. However, increased productivity does not assure an increase in employment. On the other hand, even if direct employment in agriculture does not increase, the indirect or linked employment should increase considerably. The expansion of related services and maintenance of inputs and machinery, as well as marketing are to be expected.[3]

The ratio of those employed in the services to those directly employed in agriculture may be surmised from a report on agricultural settlement in the Jordan Valley.[4] According to this study, it is possible to settle 20,560 workers or a population of 111,000 inhabitants on an area of 20,000 hectares (200 km^2). The estimated number of workers to be employed in agriculture would be 13,300; another 4,500 will be engaged in production services, and 2,760 in administration and public services. Thus, roughly one third of those employed in agriculture would be engaged in related activities. Production services, including packing, classification and transportation and distribution will of course depend on the product and the location of the market. It may be suggested, further, that the application of more efficient methods of cultivation and better utilization of inputs will allow an increase in the number of people employed in agriculture by about 35 per cent.

The investigation of water resources in the West Bank and Gaza Strip shows an enormous gap between water consumption at present and the potential in the future. The consumption patterns in the West Bank and Gaza vary widely. Water consumption is about 100 million m^3 a year in each of them, in spite of the big differences in territory and land-under cultivation. While there may be overuse of water resources in Gaza, there is a vast under-utilization in the West Bank.

The water potential is related to the hydrogeographical network and the various categories of soil composition. Obviously, the level of rainfall is one of the main factors influencing the water potential. There is very little information dealing with water potential in the West Bank and Gaza. No research has been undertaken in this field, nor have exact measurements of water consumption been established. The estimated water consumption of about 100 million m^3/year depends on the simplest methods of water use. There have been no essential changes in the use of water resources during the past century.

Rainfall in the West Bank varies from 700 mm per year in Ramallah to 100 mm per year at the Dead Sea. In the mountainous region of the West Bank, the average rainfall is 600-700 mm per year on the west side of the mountains, and 450 mm, decreasing to 250 mm, in the valley. The average rainfall in this entire region is estimated at 600 mm a year. The rainy season is 4 to 5 months. The total quantity of rainfall is estimated to be 700 million m³ a year, which is partly absorbed into the ground, filling the sources and wells; the remainder becomes run-off water. The run-off water to the west is absorbed by the rivers and sources of the Israeli territory, while the run-off rainfall to the east is consumed by the hot and deep springs in the Jordan Valley. It has been estimated that the water potential of the Jordan Valley is approximately 100 million m³/year while 50 million m³/year can be accumulated in specially constructed water reservoirs. There also exists 50 million m³ of water in the Ein Feshka area (Dead Sea coast), which contains some salt. It may be suggested that the Jordan Valley may be able to utilize additional water through joint projects with neighbouring countries.[5]

The present consumption of water in the West Bank consists of approximately 100 million m³/year: 43 million m³ in the Jordan Valley, 50 million m³ in the mountainous region for use in irrigation; and 7 million m³ in household consumption and for industry in the towns.

The recent increase of consumption in the towns and large villages, which is due to development works being carried out by the authorities within recent years, may be seen as proof of the possibilities to increase the water potential through investment in the search for additional water resources. Some examples of the increased water supply may be helpful. This Hebron-Jerusalem-Ramallah region had a capacity of 145 m³ an hour in 1967. Today, it has increased its capacity to 545 m³ an hour by erecting regional water supply systems, to utilize the water in a rational manner. The water supply of the district of Ramallah north of Jerusalem, was based on the distribution of 100 m³ from local resources and 50 m³ received from East Jerusalem. Today, the supply has been increased to 200 m³ from local sources only. In Nablus, the Jordanian administration carried out drillings for the supply of 100 m³ per hour. Today the capacity has increased to 300 m³ an hour although only 100 m³ are consumed because of the shortage of electric power.

It should be noted that the limited water consumption is a result not only of lack of available sources but of the high costs of supplying water to the mountainous region: 4 to 5 IL per m³ while in Israel, home consumption is about 1 IL per m³. One of the reasons for this difference is that the price of water in Israel is subsidized. The price is based on the cost of production, therefore, water consumption in the West Bank towns is only 20 m³/year while in Israel it is 80 m³/year per person.

Some estimations for the future consider 26-30 m³ a year per inhabitant as acceptable. We will use the quantity of 50 m^3 per year per inhabitant as a norm in our study. Such a norm means a supply of 50 million m^3 per year will be sufficient for an additional population of one million. If the future development were based on urban and industrial development patterns, then adequate supply would be available not only for one more million but for an additional 3 to 4 million people simply by more efficient use of the existing water resources.

Noteworthy are the existing possibilities of increasing water supply by desalination, both for agriculture and for human consumption. The research currently undertaken on the cost of desalination of water for human and industrial consumption in the Negev, shows that the economic cost may not be a hampering factor for development because it is not more than 2 per cent of the GNP per capita. However, desalination is more convenient and less expensive for the region of the Gaza Strip than for the West Bank. In the latter, water supply may be doubled and even tripled simply by more efficient use than has been the case.

In the Gaza Strip, the average rainfall is 400 mm, but it rapidly decreases in the south where, in Rafah it is about 150 mm a year. The main water supply is from the groundwater reserves which are used very intensively. The water is generally drilled from underground sources and wells, supplying a total of about 100 million m^3 annually. This water is mostly used for agriculture and partly for the large urban population. Lately there has been a considerable amount of over-pumping and there is danger of increasing salinity of the water. Successful experiments were carried out to supply desalinized water to this region.[6]

To summarize, the comparison between the potential and actual consumption of water sources shows that at least 400 million m^3/year may be available in the West Bank to supply not only the requirements for future urban-industrial development but also to irrigate a substantial part of the cultivated area. We foresee no problems in ensuring water for urban-industrial development for a population of three to four million, which will require 200 to 300 million m^3/year, even if we depend on the known resources only. New technologies and water sources would make the prospects more optimistic.

The increase of agricultural output and income as water supply increases will depend on the amount of additional water available to agriculture and the efficiency of its utilization. For example, the scarce water resources may be devoted to high income crops; more efficient utilization of water is possible. For example, 4,000 cubic meters are used in the Jordan Valley for each dunam of citrus, while only 1,000 cubic meters per dunam are used in Israel for the same crop.

Efficient use of the resources also reduces the working time per unit of land and unit of output, but it increases the working time in the related services. Therefore, the rate of increase of employment in agriculture is considerably lower than the forecasted increase in agricultural output. At the same time, mechanization and other methods of modern agro-techniques are at present difficult to introduce because of the land tenure system. The structure of land ownership, the high frequency of small farms, and the enormous fragmentation of the arable land even in the small and medium size farms make mechanization difficult and thereby reduce the effects of modern technology in agriculture. Consolidation of the existing farms and the creation of large-scale farming are therefore basic conditions for future efficient land use. A programme of agrarian reform would enhance these efficiencies and raise productivity in agriculture.

The progress of agricultural output depends on the simultaneous application of these different methods and techniques. The experience of the last few years, even in unfavourable conditions, makes it seem reasonable to conclude that agriculture will be able to produce enough to cope with the increase in demand, and will provide a per capita income comparable to other sectors of the economy.

There is a great difficulty in forecasting the figure for additional employment in agriculture. As has been mentioned, technological progress influences agricultural output, but not necessarily at the same rate or in the same direction as it affects employment. However, the extension of irrigated areas as a result of additional use of water may increase the employment figure in some high income crops, especially vegetables which require a substantial increase of manpower.

If we assume that an additional 400 million m³ of water per year can be derived from the Jordan Valley resources and other projects constructed in co-operation with neighbouring countries the number of those directly employed in agriculture may be increased by an additional 55,000 workers.[7]

Another way of forecasting agricultural employment is to consider a feasible remunerative farm to be about 2.5 ha (25 dunams) at least half of it irrigated. The increased effective water supply will make it possible to generate much more employment out of the presently cultivated 200,000 ha. At an average of 2.5 ha per farm we will have the equivalent of 80,000 such farms, employing at least 80,000 workers. If we can assume that additional land will be brought under cultivation the number may be raised to about 100,000 workers.

The Contribution of Industry

The survey of land potential for future development and absorption of additional population indicates that a space of about 2,300 km² is available for utilization; of these 100 km² are in the Gaza region and the rest in the West Bank. The land quality is similar to the land presently utilized. Of the total area of the West Bank and the Gaza Strip only about 1,000 km² are considered unusable for human settlement.

The ability to exploit this space for the absorption of additional population depends on whether it is used for agro-rural or for urban-industrial development. The latter is the major absorber of people, as there is a big difference in how the land will be used in the new settlement schemes. At present, about 2,000 km² in the West Bank serve a rural population of 400,000, or about 5,000 m² per person. In the Gaza Strip where 45 per cent of the agricultural land is irrigated, 200 km² sustain a rural population of 60,000 people or about 3,330 m² per person. In contrast, the survey of land use in urban Hebron shows that only 150 m² are used per person. This figure may actually be too small for all the urban uses, assuming a modern quality of urban life. The difference of land allocation between the agro-rural and the urban-industrial uses is common to all other countries.

To take an example, the Netherlands may be good for comparison. The surface of the Netherlands is 40,000 km², of which 5.3 per cent or 2,145 km² is used in urban settlement. The average space in urban use for the urban population of 8 million is 268 m² per person. Because most of the urban settlements are small, the land surface per inhabitant is relatively high. The Netherlands' development scheme for the year 2000 is based on a forecast of 20 million inhabitants, of whom 14 million will live in urban settlements. The surface allocated for this population is 11 per cent of the total surface or 4,500 km² which gives 320 m² per person. The aim of the planners is to increase the percentage of single family homes from the present 30-40 per cent to 50-70 per cent of total housing in the year 2000. In contrast, the allocation of land space for agriculture is estimated at 2,500 m² per person.

Our assessment of the future development will be based on industrial development and urban living. We shall explore also the potential to create employment, the investment needs and the marketing potential or industrial goods.

To get a clear picture of the nature and magnitude of land require-ments for the future development of the Palestinian state, it may be useful to look at some international estimates.[8] There are great differences in the estimates of land requirements according to the different uses. Sometimes the differences are the result of topographical conditions. In particular, densities in residential areas (high percentage of one-family or multi-storey houses) and also the rate of motorization influence the land requirement for roads. According to the calculation made for planned settlements, land requirements in the European countries are estimated as in Table 6.2.

Table 6.2: Use of Urban Space

	Average m^2/person
Residential	110
Roads	40
Green space	48
Public services	30
Industry	30
Commercial services	12
Total	270

This allocation illustrates the impact of planned land use on the land requirements. Actual land use norms are usually higher than the planned land use norms.

In an economically less-developed country with a low degree of motorization and with a high percentage of the population employed in agriculture, a larger percentage of space may be necessary as residential area for widely spread rural settlements. Generally, less land will be needed for larger human settlements where the densities may be higher than for smaller towns. The above estimate of 270 m^2 per person is closer to the maximum as a national planning norm.

For our estimate we shall use the same figures for housing as in the developed countries, but with a slight reduction of the allocation for roads and green spaces. However, our allocation will be considerably higher than allocations in the developing countries. We will use the same norms for industry and for public and commercial services as in the developed countries. Our allocation is based on the assumption that a high percentage of the labour force will be employed in industry and services and will use modern technology. The high norms for housing are based on the sociological characteristics of the people, their rural origin, their interest in living in single family homes rather than to be located in multi-storey buildings. Table 6.3 summarizes our estimates of the urban land allocations for future development in the West Bank and Gaza Strip.

Table 6.3: Projected Urban Land Allocation

	West Bank & Gaza Strip m^2/person	European Countries m^2/person
Residential	110	110
Roads	25	40
Green space	20	48
Public services	30	30
Industry	30	30
Commercial Services	12	12
Total	227	270

According to these data, a space of 227 km^2 will be sufficient to settle a population of one million inhabitants within the framework of urban settlements and industrial occupations.

The estimation of the total land requirement for future development should take into account the additional space needed for regional and national land requirements, such as roads, recreation and power stations. This item is especially important considering that a high percentage of the additional population will probably be interested in settling in small or medium-sized urban settlements according to the respective place of origin. Obviously a large number of small urban settlements would require more land allocation for regional and national purposes than would be required by a few big urban agglomerations of population. We will therefore use the same norm of land allocation of regional and national purposes as for urban settlements.

According to this assumption, an additional 227 m^2 per person may be required for regional and national purposes, or a total of 454 m^2 per person, and 454 km^2 will be required for each additional one million people. Given the potentially available area of 2,300 km^2 in the West

Bank and Gaza, it is evident that about 5 million more people can be settled in the new state, and many more if the land presently in use is utilized more efficiently than it has been.

The future industrial development will depend in part on the size of the internal market, which is a function of the level of income and the size of the population. It also depends on the export market which may be fairly extensive, given the prospects for high capital investment and the growing consumer demand in the Middle East countries and in countries outside the region. It is also highly probable that the export market will be greatly be enhanced by the potential supply of highly qualified manpower among the Palestinians, large capital investment from foreign sources, and by the fact of independence that would be enjoyed by the new state.

It may be of particular interest to look at a study of the export market conducted in the context of a plan to develop the administered territories by Israel.[9] This study, however, was based on an assumption of close economic ties with Israel. It was carried out in a period when some intellectuals and political leaders were convinced that strengthening the economic ties with the occupied territories and that solution of the refugee problem would create favourable conditions for solving the Arab Israeli conflict. The study had three aims: (1) to assess the feasibility of erecting industrial towns near the 1967 borders; (2) to develop methods of assessing the prospects of industrial towns; and (3) to prepare a plan of the infrastructure for creating industrial towns.

For purpose of this study, an industrial town is an urban structure based on four central principles:

a. Industrial exclusiveness.
b. Distance from residential areas.
c. Growth possibilities.
d. Comprehensive planning.

The first principle means that an industrial town is designated solely for industrial plants and all the services necessary for these plants. An industrial town does not include residential areas for its workers. This means the delivery of the means of production to the natural residence of the workers (agricultural and rural areas, small towns, etc.) and not vice versa, as is nowadays customary in all industrial towns.

The meaning of the second principle is clear: the industrial town should be located in close proximity to existing residential areas, so it would not become an industrial suburb of an existing city. The determination of the desirable distance of an industrial town from its workers' residence depends on many factors: available space, land conditions, ecological factors, traffic routes, etc. On the one hand, the principle of detachment of the industrial town from residential areas has to be kept. On the other hand, the need to transport the workers over large distances should be avoided.

Several locations were found suitable for industrial towns in the West Bank and Gaza, and many products were considered good prospects. The locations were widespread with no less than eight points regarded suitable. It was estimated that 200 m^2 per worker would be sufficient for the enterprise; that is, 2.5 to 20 km^2 would be needed for the town that would employ between 15,000 and 100,000 workers. The proposed commodities were selected from the Index of Standard International Trade Classification. The criteria for selection included availability of raw material, manpower, transportation facilities, distribution and marketing institutions, and the export prospects to Israel, the Arab countries and OECD. The final choice of commodities depended on the needed investment per worker, the expected output per worker, and the expected profit. The land required was found to be a function of the specific industry, as shown in Table 6.4.

One hundred and twenty-five products were chosen, of which thirty-five commodities were recommended for the first stage of development. The long list of commodities considered feasible for both the internal and external markets ranges from textiles and crafts to the manufacturing and assembly of cars and machinery.[10]

Table 6.4: Proposed Industries and Area per Worker in each Branch

Branch	Built Area per Worker in sq. m.
Electricity and Electronics	30
Textiles	31
Food	42
Metal	52
Wood	60
Rubber and Plastic	63
Average	46.3

In terms of 1975 prices, the required investment per worker varied from $1,000 in textiles and leather industries to $50,000 in the tobacco, cement, clay and lime industries. The average investment per worker was \$12,000, of which \$7,500 would be in infrastructure and \$4,500 in machinery. A reasonable estimate of the required investment in 1977 might be around \$18,000 per worker. The specific locations and commodities, however, must be determined in due time by the planners in the state of Palestine, according to the conditions then prevailing.

D. The Contribution of the Building Industry

It may be estimated that the requirements to improve the housing conditions of the present population and settle the refugees will increase the number of employed in the building industry by no less than 25,000 workers. There is a great difficulty, however, in estimating the demand for manpower in establishing towns for the additional population of over one million people. The number of new jobs will depend on the allocated space per person and the rate at which resettlement will take place. If we were to use the average of 10 m^2 per person or 60 m^2 for a family of six, 10 km^2 of built area would be needed for each additional one million people. An additional 25 per cent of the built area would be required for public and private services. The total required built area might be estimated at 12.5 km^2 to 15 km^2 for each one million people.

We estimate the required manpower by measuring the output per worker in terms of completed built area (buildings begun and buildings completed). In 1975, a labour force of 10,000 in construction in the West Bank and the Gaza Strip completed the building of 558,000 m^2 and began the construction of another 778,000 m^2, or an average of 668,000 m^2 in one year. According to this estimate, the output per worker is 66.8 m^2 in a year. This output may be an overestimate since in actuality more manpower was employed in the building activity than the registered 10,000 workers in the form of unpaid labour. Assuming this unpaid labour amounts to 20 per cent, the output of 55 m^2 per worker year would be closer to reality.

Assuming the resettlement of all the returning refugees can be accomplished within five years, the yearly constructed space would be between 2.5 to 3 million square meters. The number of workers required for such an accomplishment would come to 45,000-55,000 per year. Adding the number of workers required to improve the housing conditions of the present population of the West Bank and Gaza would bring the number of new jobs to 75,000 during the first five years.

In a country which has pledged itself to double the population within a short period of time, a large portion of the resources will be required for carrying out that pledge. The building of infrastructure, roads, water supply, sewage, power stations, irrigation works, etc., would require additional manpower. An estimate of an additional one third of the construction labour force may be needed, or 25,000 workers. The total manpower required in the building industry would then come to about 100,000 workers, inclusive of the 10,000 presently employed, each year.

This assessment, while it may seem unrealistic, is consistent with the experiences of other countries with high immigration rates. The building industry has usually been the main source of employment during the early stages of settlement and development. It has also been a major force in determining the growth pattern of the economy.

The Contribution of the Service Industry

The expected increase of population and a higher GNP per capita should create a relatively high demand for services. The modernization activities in agriculture and industry will probably necessitate the creation of a network of economic services capable of meeting the new demand. In addition, the creation of an independent state would require an increased number of public servants, and a network of social services such as education, health, banking and communication. The development of tourism, especially when peace reigns in the area, will be an additional source of employment. These services may be directly related to the infrastructure. The level of physical and social infrastructure might be seen as one of the deciding factors influencing socio-economic development.

The erection of an appropriate infrastructure is critical in achieving a high rate of economic growth concurrently with the influx of a large additional population within a short time period. The required investments in infrastructure include the primary development works such as roads, water, sewage, energy sources, and the secondary social infrastructure including education, health, etc.

The costs of infrastructure works include investment in the local and regional services, the costs of erecting human settlements for an additional population, and for improving the conditions of the permanent present population. At the same time, investments required for infrastructure on the national level such as development of water and energy resources, national roads, port and airport facilities, should be estimated. Our estimates of the investment requirements for an additional population are based on a comparison with investment in human settlement developments in different countries.[11]

According to these international data the costs of infrastructure works for an additional population in urban settlements per capita are as follows: in Turkey, US$1,000; in Ireland, US$1,910; in the UK, US$4,500; and in Norway, US$5,360. The GNP per capita distribution of these countries in 1972 was as follows: Turkey, US$495; Ireland, US$1,839; UK, US$2,714; Norway, US$4,071.

The large differences of the infrastructure costs between the more developed and the less developed countries are due to the level of GNP and the pattern of consumption such as the services of subways, highways, dense telephone network, recreation services, large housing space, and also the quality of buildings.

The following figures show the differences in housing costs: Turkey pays US$2,700 for the dwelling of a family of 5 or US$540 per capita; Ireland spends US$6,000 for a dwelling of a family of 5 or US$1,200 per person; the UK spends US$12,000 or US$2,335 per capita. Table 6.5 provides relevant data regarding the infrastructure expenditures.

From our estimation, we will mostly use the data from Turkey and Ireland, since the GNP level in these countries comes closer to the forecasted level in the Palestinian State than that of Norway and the UK.

The differences between Turkey and Ireland consist mainly in the expenditure on housing. A difference of US$800 in housing corresponds to the generally higher expenditure in Ireland ($1,910) than in Turkey ($1,100). This means that the expenditures on other items are similar.

The expenditure on housing in Ireland is 60 per cent of the general infrastructure costs, while in Turkey it is only 38 per cent. Probably, the ratio between the housing costs and other costs may be estimated as the average between the ratio in Turkey and the ratio in Ireland, or housing expenditures equal to 50 per cent of the total costs on the infrastructure.

The housing costs for absorbing the additional population in the West Bank and Gaza Strip will be based on the cost of housing in the West Bank which is about US$150/m^2. We take as a norm between 10 m^2 per person and 12.5 m^2/p.[12]

If we estimate the housing costs at $1,500 per person, we shall add an additional $1,500 to cover the other infrastructure costs including physical and social infrastructure. The general infrastructure cost for an additional population of one million in new human settlements should be therefore evaluated at US$3,000 million.

Table 6.5: Human Settlements Costs ($ per capita in 1968 prices)

Physical Infrastructure	Turkey	Ireland	Norway	UK
Roads	95	120	771	720
Water	36	48	–	60
Sewage	77	72	31	72
Electricity	65	60	800	188
Gas and telephone	–	–	–	65
Total	273	300	1,602	1,105
Social Services				
Education	178	290	468	235
Health and other services	–	–	–	108[a]
Social institutions	–	–	142	72
Total	178	290	610	415
Centre of the town	–	–	210	–
Parks and recreation	–	–	68	24
Transport, communications and other services[b]	230	–	–	–
Housing	412	1,200	2,600	2,335
Land	–	120	270	295
Fees and Interests	–	–	–	410
Total	1,100	1,910	5,860	4,584
Commerce	–	–	–	518
Industry	–	–	–	504
Total	–	–	–	5,606

a. A hospital for 100,000
b. This figure includes also recreation, telephone.

Source: Turkey, Ireland and Norway: Study of Council of Europe. UK: A.P.
Stone, *The Structure, Size and Costs of Urban Settlements,* Cambridge
University Press, 1973.

For evaluating total costs, $1,200 million should be added in order
to absorb the 400,000 refugees living in the outskirts of the existing
urban and rural settlements, as well as to improve the living conditions
of the permanent population in the West Bank and Gaza Strip. A
summary of the expenditure on infrastructure for one million people
is contained in Table 6.6. The actual cost estimates will be prorated

according to the size of the population to be settled each year.

The share of services in employment will probably not be drastically changed from the present 47 per cent. Our estimate is 49 per cent of the labour force or 167,000 workers out of 546,000. However, a major change may be the reduction of disguised unemployment, the reduction of under-employment, higher labour productivity and an increase of the contribution of the service industry to the total GNP. Whether these changes will take place will depend on the pattern of investment and the modernization process in that industry.

Table 6.6: Estimates of Infrastructure Costs for One Million People

Items	In $ million
1. Electricity	300
2. Roads in the West Bank	200
3. Roads in Hebron and Gaza	70
4. Gaza port	70
5. Desalting plants	200
6. Other water development projects	100
7. Natural resources development (Dead Sea)	200
8. Tourism	100
9. Education, health and public services in national framework	260
Total	1,500
Summary of investments	
1. Human settlements for one million additional population	3,000
2. Improving conditions of the existing population (prorated)	1,200
3. National infrastructure	1,500
Total	5,500

Notes

1. Guesses exceeding 150,000 have also been made.
2. This observation is supported by data on scholarships held by Palestinians in 1974-5. UN Report of the Commissioner General, *Relief and Works Agency for Palestine Refugees in the Near East*, p.75.
3. In industrialized countries, the number of employed directly in agricultural activities exceeds the number of those directly employed in agricultural production.
4. Ministry of Agriculture, Israel, 1967.
5. A quantity of 190 m^3/year from the Yarmuk and Jordan Rivers, 160 m^3

from the Yarmuk and 30 m^3 from the Jordan delineated in the Johnston Plan may be available.

6. J.F. Fried, *A North Sinai-Gaza Development Project,* Washington D.C.: Middle East Institute, 1975, Mimeo; J.F. Fried and M.C. Edlund, *Desalting Technology and Middle Eastern Agriculture,* N.Y. Praeger, 1971.

7. The number of employed in production and distribution services will be included in forecasting employment in the services sector.

8. For the most part, these details are taken from *Urban Land Policies,* prepared by H.Darin-Drabkin for the United Nations, 1973, and from his forthcoming book, *Land Policy and Urban Growth.*

9. Y. Orgler, *Establishment of Industrial Towns Near the Green Line,* Tel Aviv: Cherikover, 1973.

10. The list of commodities, with estimates of output, investment and expected markets is attached as Appendix 1.

11. Data from the UN and from the Council of Europe, 1968.

12. Such a norm seems quite low when compared to the industrialized countries. however, it is very high when compared to the developing countries. For instance, in India, the Bombay Development Scheme attempts to provide a dwelling of 20-50 m^2 for a family of 6 persons. In Singapore, where the GNP is $700/p, the norm is the same.

7 THE VIABILITY PROSPECTS II – THE OVERALL PICTURE

We are now in a position to pull the various sectors together and evaluate their viability and internal consistency as parts of the economy as a whole. We shall also look at the human side, as the subjective approach to viability.

The Aggregate Economy

Several assumptions have been made as follows:

1. Population will continue to grow at a relatively high rate during the early years of statehood which we regard as the transition period. We have assumed a population growth rate of 2.5 per cent annually, although this rate may be high, in which case it will bias the results against viability and make the test more severe.

2. A transition period of five years will be necessary for the absorption of the returning Palestinians. The number to return each year will be a fraction of the population remaining to be repatriated, such that by the end of the fifth year all those expected to return will have been repatriated. Specifically they will return as follows: one fifth by the end of the first year; one fourth of those remaining by the end of the second year; one third of the remaining by the end of the third year; one half of the remaining by the end of the fourth year and the rest by the end of the fifth year.

3. The participation rate in the labour force will be 23 per cent of the population, which is somewhere between the present low rate of 18.6 per cent in the West Bank and Gaza and the much higher rate in more advanced economies.

4. Employment will be concentrated in the services and in residential construction during the first few years; it will be assumed that 49 per cent will be in services, and as many as needed to cope with the demand for residential construction, at the rate of one worker per each of 55 square meters of built space. The total dwelling space will be determined at the rate of 12.5 square meters per capita for the returning refugees and the 400,000 refugees in need of housing in Gaza and the West Bank. Employment in agriculture has been assessed on the assumption that water supply will increase through higher efficiency, more drilling and conservation, and through desalination. Taking as an

average work year to be 256 days, and an average number of work days per unit of land with a variety of crops common to the area, we estimate 25 dunams (2.5 hectares) per worker to be sufficient.[1] Given the availability of 2 million dunams presently in cultivation (200,000 hectares), 80,000 workers may be absorbed. Should additional land be put into cultivation, more workers may then be added into agriculture. In the meantime, the additional labour force to be used in agriculture will be fixed and will be distributed in equal numbers over the five-year transition period. Finally, employment in industry, which includes industrial construction, will be the outlet for the residual labour force, to achieve full employment.

5. The projected annual income per capita is $800, at the 1975 prices, during the transition period.

6. To estimate the GNP during the transition years, we have used the labour output ratio of 1974-5 as a basis.[2] The estimates have been converted into US dollars at an average exchange rate for that year of US$1 = IL 6.32.

7. The required capital for the transition period has been estimated on the basis of the costs of residential construction and employment creation in each of the basic four sectors. These estimates have been made with the help of comparative data from Ireland, Turkey, Israel and the West Bank and Gaza.

8. All values have been calculated at the 1975 prices, although the year 1977 has been taken as the point of departure and the year 1978 as the first year of transition.

9. The most difficult problem has been to determine the sources of capital to finance such a large influx of people within the transition period. Some heroic assumptions have been made regarding the role of compensation for the lost property to be paid by Israel, and the contribution of countries outside the region. The most critical assumption is that an Arab Fund for the State of Palestine will cover the deficit in capital requirements by arrangements to be decided by the parties concerned.

10. Finally, in order to facilitate economic integration of the returning people with the permanently established people, and to prevent duality from becoming a serious issue, an allocation has been made to improve the conditions of the present permanent residents of the West Bank and Gaza.

The emerging economy is described in Tables 7.1-7.4. As shown in Table 7.1, the population will increase from the 1975 figure of 1,112,000 in the West Bank and Gaza to a total of 2,821,937 in the

Table 7.1: Population, Labour and Output in the Transition Period

Year	(1) Population	(2) Pop. Increment	(3) Labour[a] Force	(4) Lab. Force Increment	(5) Expected[b] GNP $000's	(6) Exp. GNP/Wrkr. $
1978	1,469,309	301,014	337,941	127,561	1,175,447	3,477
1979	1,784,644	315,335	410,462	72,527	1,427,715	3,477
1980	2,111,972	327,328	485,753	75,285	1,689,577	3,477
1981	2,458,940	346,968	565,556	79,803	1,967,152	3,477
1982	2,821,937	362,997	649,045	83,489	2,257,349	3,477

a. Based on a 23 per cent labour force participation rate.

b. Assuming $800/capita income

Palestinian State by the end of the transition period. The labour force will make a big jump during the first year, increasing by 122,561, but stabilizes around an increment of 75,000 workers a year during the rest of the transition period. Given the projected income per capita and the expected labour/output ratio or GNP/worker ratio, the GNP in the last transition year will be about $1.9 billion or about $300,000 million short of the target, assuming full employment. However, if we assume a rise in productivity of labour over the transition period of 10 per cent, the gap will vanish. The same will happen if we assume that the new investment, especially in industry, will be at a higher level of technology and productivity than had existed in the traditional economy. Therefore, as far as labour productivity is concerned, the projected output in the new state will meet the expected per capita income.

Table 7.2 shows the distribution of employment and the projected income by sector. As may be seen, employment in agriculture does not change dramatically. The main source of employment in the early years will be in the services, public and private, and in residential construction; together they would absorb more than 60 per cent of the labour force by the end of the transition period. A dramatic change will take place in industrial employment as it will multiply five-fold between the first year and the last, as expected. The shift of emphasis to industry will become more evident as the residential construction begins to stablilize after the transition period is completed.

Table 7.2: Employment and Income Structure in the Transition Period

| | (1) Agriculture | | (2) Industry | | (3) Construction | | (4) Services | | Total Output |
	Employ. Workers	Output[a] $	Employ. Workers	Output[a] $ m	Employ. Workers	Output[a] $ m	Employ. Workers	Output[a] $ mil.	$ m
1978	63,200	175.0	22,556	36.0	86,594	430.6	165,591	470.9	1,112.5
1979	72,400	200.5	47,091	75.1	89,848	446.8	201,129	572.0	1,304.4
1980	81,600	225.9	73,381	116.2	92,574	460.3	238,018	676.9	1,479.3
1981	90,800	251.4	100,596	160.4	97,038	482.6	277,122	788.1	1,682.5
1982	100,000	276.4	130,332	207.7	100,681	400.7	318,032	904.4	1,889.7

a. Output at 1975 productivity level. No allowance for growth in productivity has been made. Estimates used are those of the West Bank, converted into dollars at a rate of $1 = IL6.32, are as follows: Agr. = $2769 per worker; $1594 in industry; $4973 in construction; & $2844 in the services. Computed from Bregman's data, p.39.

Table 7.3: Additional Investment Needed for Housing and Employment Creation[a]

Year	(1) Housing	(2) Agric.	(3) Indus.	(4) Resid. Const. $ MILLIONS	(5) Services[b]	(6) Total	(7) Total[c]
1978	714.4	128.8	30.7	229.8	533.4	1,637.1	1,837.1
1979	741.3	128.8	294.14	9.8	213.2	1,387.5	1,587.5
1980	763.7	128.8	315.5	8.2	221.3	1,437.3	1,637.3
1981	800.6	128.8	326.6	13.4	234.0	1,503.4	1,703.4
1982	830.6	128.8	356.8	10.9	245.5	1,572.6	1,772.6
	3,850.0	644.0	1,324.0	272.1	1,447.4	7,537.9	8,537.9

a. At 1975 prices.
b. Public and private.
c. Column (6) + 200 per year to improve the conditions of the permanent residents.

Agriculture: incremental employment = potential employment — 1977 estimates prorated over a five year period: $\frac{100,000 - 54,000}{5} = 9,200$ workers a year @ $14,000 per worker. Industry: employment is a residual of the labour force @ $12,000 investment per worker. Residential construction: required housing (12.5 m² per person) @ 1 worker for each 55 m² @ $3,000 per worker. Services: Employment equals 49% of labour force at an investment of $6,000 per worker. Investment requirements per worker are averages derived from Table A1.

Housing: investment equals population requirements at the rate of 12.5 m² per person, @ $150 per square meter, as explained in the text.

A slight double counting may have occurred between columns (1) and (6) but not enough to affect the conclusions.

Table 7.3 shows the capital requirements for the absorption of the returning population and the refugees still living in camps in Gaza and the West Bank, and capital for improving the conditions of the rest of the population. These requirements are shown on an annual basis. Again we notice a large sum required in the first year, compared with the other years, because of a major discontinuity in the economic structure and the necessity of creating a large number of new jobs. After that a certain degree of stability may be noted, in part because additional investment will be needed to finance the incremental job creation only. This is possible on the assumption that investments made in the transition period will still be good at the end of the period. If we were to allow for a depreciation rate of 10 per cent annually, the gross investment requirement will have to be raised by that rate.

It may be significant that the amount of capital needed is less than $9 billion in 1975 prices. Even if we raise the figure by 50 per cent, to allow for inflation, the total sum will be less than $13 billion to be dispensed over a five-year period to absorb the newcomers and put the state on a sound basis.

Two main questions should be raised at this point: Where will the capital come from, and what industries should be established to absorb the labour force, outside the services and residential construction? The issue of which industries has been discussed above and we have little more information to permit detailed analysis or more specific recommendations at this stage. Feasibility studies within the framework of specified trade relations and a known political system will be necessary to determine the appropriate industries. Table 7.4, however, provides an illustrative answer to the question of capital sources. We have assumed that UNWRA, which has aided the Palestinian refugees since 1950, will continue to be active during the transition period. An estimate of $150 million has been made as the contribution of that agency, in addition to the invaluable role UNRWA can play in the transition. Another assumption has been made regarding domestic capital. At the present time a rate of saving of over 15 per cent prevails in the West Bank and Gaza. Therefore, we have assumed a 15 per cent rate of saving as the domestic contribution to the expansion. Compensation funds from Israel are bound to come. The extent of the compensation for about four million dunams (400,000 hectares) outside the Negev plus improvements is difficult to estimate and may be subject to negotiation.[3] It is our hunch that the total sum of not less than $3 billion will be expected; we have divided this sum to be dispensed in equal portions over the five-year transition period. We have also made

Table 7.4: Sources of Capital[a]

	1978	1979	1980 $m	1981	1982
Domestic savings	176.3	214.1	253.4	295.0	335.6
Compensation from Israel[b]	600.0	600.0	600.0	600.0	600.0
Arab Fund for Palestine[c]	480.9	423.4	353.9	377.4	402.0
Foreign aid	500.0	200.0	200.0	200.0	200.0
UNRWA funds	150.0	150.0	150.0	150.0	150.0
	1,907.2	1,587.5	1,637.3	1,703.4	1,772.6

a. At 1975 prices.
b. Amount proposed as a possibility.
c. It is assumed that an Arab Fund would be set up and would carry the responsibility to cover the deficit between these other contributions and the proposed budget.

an assumption regarding foreign aid from outside the region. While we know of no standing commitments by any country, we are confident that some contributions will be made by the United States, Germany, France, Britain, Japan, the USSR and China among others. Since the first year will be the most demanding, we have made the first year contribution the largest, $500 million, to be followed by a $200 million contribution in each of the next four years.

Finally, we regard the Arab interest in helping the Palestinians to establish a home and a state to be genuine. Accordingly, we have left the Arab countries with the responsibility to cover the deficit in the national budget suggested above. The Arab contribution would not reach $500 million in any one year of transition. We believe this sum to be feasible, especially if an Arab Fund for the State of Palestine would be created to manage that contribution, and to determine the methods by which capital will be recruited and dispensed.

On the assumption that the above requirements will be met, we may conclude that the objective criteria of economic viability can be satisfied, and a State of Palestine is economically feasible as outlined.

The Human Question, or the Subjective Conditions

One of the advantages small nation-states have is their potential for homogeneity, ease of communications, manageability and hence unity of .purpose and direction. Even when differences in goals and policies exist, it may be easier in a relatively small state to iron out the differences, make compromises, and reach an agreement to achieve unity. The experiences of the recent decade, however, tend to cast doubt on the validity of these assumptions. We need only to look at Ireland, Cyprus, Lebanon and Israel to become sceptical. Integration of the people in all these countries into a unified harmonious community has not been evident.[4] At the same time, the small nation has the disadvantage of being too small to allow regional or communal autonomy within the country as a means of unity. Therefore, it needs to be shown that the Palestinian State would be internally integrated enough to achieve viability, especially in view of its smallness, newness and the hostile environment it is bound to encounter.

It is granted that the Palestinians are all of them Arabs; they use the same language; they share the same history; they share the cultural heritage; they mostly have the same religion; and they have been subjected to the same catastrophies and suffering of the last half century. They also share the goal of having a Palestinian state of their own, in addition to the unifying force inherent in their struggle against Israel. Nevertheless, the problems that will face the state may outweigh these unifying factors and lead to disharmony that would jeopardize economic viability.

The need for internal integration arises from the conditions which have disrupted traditional harmony and resulted in dispersing the Palestinians, segmenting them, and making contact between them virtually impossible. Since 1948 the Palestinians have lived under the rule of more than ten different countries.[5] The political and social disparity between the various groups of Palestinians in the diaspora has been matched by economic inequality and diverse conditions and opportunities available to them. The Palestinians who have been living in refugee camps have had little to do with those who have integrated themselves in the economic life of the country they have resided in, such as Lebanon, Kuwait or Syria. The Palestinian Arabs who have continued to live in Israel had lost contact with other Palestinians until 1967 when they resumed a restricted contact with the West Bank and Gaza under the yoke of Israeli occupation. These last three groups have largely escaped the refugee status, the abandonment of home and land and family and friends. They may not be able to comprehend the impact of camp life on a whole new generation born there.

The educational systems and health services which the Palestinians have utilized have also varied widely, from the UNRWA services in the camps to the almost universal health and educational services in Israel and the oil-producing countries of the Middle East. In addition, there is a large cadre of professionals who have acquired training in advanced countries, but who have also acquired tastes and habits that differ much from those of their compatriots in the Middle East. The variation may, of course, be advantageous and enriching, but it has to be reconciled with the national aims and expectations of the emerging Palestinian state. This reconciliation is exactly what internal integration is all about: it is the unification of the society for the purpose of economic achievement and viability. It is the creation of the confidence of identity from which the national expectations and achievement processes emanate. And for the Palestinian State this integration is indispensable. The questions to be answered then are: once the conflict between Israel and the Arabs has been contained, will internal integration be possible at a rate favourable to economic viability? What are the prerequisites for such integration, and what are the costs?[6]

Integral integration could, on a long-term basis, come about through the market mechanism. Mobility in response to market demand would tend toward equalizing prices and wages for similar commodities and services. Presumably the market mechanism, through prices and wages, would also guide the economy toward satisfying the consumer demand, thus bringing about an equilibrium which may be consistent with the national objectives in a market economy. However, in the newly created Palestinian economy, the market mechanism is bound to be underdeveloped for some time; serious short-run problems of development, employment and income will be crying for a solution at a rate faster than may be expected from the market mechanism. Planning, therefore, may be the only means to assure integration of the economy and develop the market rapidly enough to assure viability. Planning, however, implies unity of purpose, agreement on methods, and the expectation that the plans will be implemented because they are consistent with the national objectives.

Unfortunately, there are no studies of the impact of dispersion on the Palestinian identity, unity of economic purpose, or feasibility of co-ordinating and implementing an integrative economic plan. The only known study is one of attitudes and perceptions in which the Palestinian identity seems to be associated with the 'struggle identity'. As summarized by its author, the study suggests that

deprived of a normal development in Palestine, camp Palestinians have constructed for themselves a role in the wider Arab area, compounded out of insistence on a more complete independence, cultural loyalism and modern skills in construction, technology and science.

Of course this amalgam of qualities and aims is very far from being a fixed identity; on the contrary, it is a fluctuating thing, and each Palestinian expresses it in a different way. But even if official organizations like the PLO fall into confusion, Palestinianism exists, and has effects, because the Arab world has neither been able to wash its hands of their problem nor to solve it.

Whether or not this still-emerging Palestinianism crystallizes around a state, or a movement, it is likely to play an important, even if unpredictable, role in the ever changing reality of Arab politics and culture.[7]

There are few other studies of countries or communities that may be generalized to the Palestinians. A study that may have some relevance deals with the Austrian society after World War II. The experiences of Austria indicate that a major effort was necessary to bring about a sense of national identity, starting with grade school education and reaching all the way to national economic policy and implementation. But it could be done in a relatively short order. As recently summarized, The Austrian experience was as follows:

Elementary teachers were instructed to inculcate pride in being Austrian not only by displaying to their students the monuments of past greatness but also by pointing out the accomplishments of the Austrian present.

Repairing war damage, building new housing for the people, new bridges, factories, hydro-electric facilities, schools, churches, and other public edifices; increasing the people's standard of living; the accomplishment of workers and farmers, of the trades and commercial people, of the Austrian public administration; the successes of Austrian scholars, artists and technicians, including those abroad; the work of the various social services – all these things are not only excellent material for patriotic instruction. . .in the third and fourth years of the elementary school but also furnish useful themes for many of the specialized programs of the advanced classes as well.[8]

Most other cases of national integration have involved different races, colours, religions, languages, or a combination of these. Success in these situations has been relatively small and can hardly be learned from, except possibly to discourage attempts to integrate. Economic domination or exploitation of one group by another has been the common result of such interaction.[9] The only other case studies we may point to as relatively successful are those of the United States which took a long-term perspective through the market mechanism, and the Soviet Union which resorted to planning and a whole system of indoctrination and re-education. These, however, are large economies which can tolerate market segmentation, variety of economic objectives, and even a certain degree of autonomy of sub-groups, as happens to be the case with the individual states in these nations.[10]

It may be ironic that the most relevant example in this context is the state of Israel. The Israeli Jews, like the Palestinian Arabs, share the racial origins, the language – to an extent – the religion, to an extent, the diversity of experience because of where they had lived for a long time, and the sense of purpose in returning to a homeland. The Palestinian Arabs, however, have an advantage in that their diaspora has not been as long and their diversity not as serious and thoroughgoing – there is no such difference among the Palestinians as there is between Moroccan Jews and German Jews, or between Yemeni Jews and American or Russian Jews. Therefore, if we accept the assumption that internal integration in Israel has been moderately successful, we should expect that integration of the Palestinians and the development of the confidence of identity to be at least equally feasible within a short-run horizon. The issue is how to bring about such integration at the least cost to the various groups in the new state?

The Palestinian situation has a certain uniqueness to it which may work in favour of integrating the society and the economy at a relatively low cost. The Palestinians have been faced with many trying situations and have so far managed to cope with them. They have, for the most part, been exposed to urban conditions and occupations, whether as professionals or as unskilled workers. They have broken away from traditional subsistence agriculture. They have also been exposed to new technologies which should serve them well in the development of the new state.

The Palestinians, those who have been dislocated and those who have retained their homes, are anxious to establish their own national economy. They have ideas of what they would like to do and how to go about doing them, albeit no blueprint or systematic procedure is known to exist. Thus, they have the incentive and the urge to achieve. Therefore, the development toward economic viability should permit these people to put their ideas into action. While central planning may be unavoidable at the beginning, grass-roots participation would be indispensable in building the plan and guiding its implementation. Advance surveying of interests and abilities, periodic consultation with the people, and response to the demands of these people would ensure their positive participation and facilitate their integration.

Whether returning from refugee camps or from luxury apartments in other countries, most of the Palestinians will be in need of homes, educational facilities, health services and jobs. These basic needs could be the focus of attention in the new state as an integrative mechanism. While these industries and services will not earn foreign currency and may not greatly enhance the industrialization of the economy, they are strategic in creating viability: they utilize local resources to a large extent; they provide employment; they provide food and shelter and other services to the people who may have been deprived for a long time; they provide tangible evidence of achievement; and, most important of all, to concentrate on these industries would show the returning Palestinians that their needs head the list of priorities in the new economy, as these industries would mean resettlement and rehabilitation in a national home.

The planning of resettlement and rehabilitation offers another advantage: it provides the opportunity to plan new towns in ways that are in harmony with the wishes of the prospective inhabitants; it would allow them to form their own communities, and build their own homes, rather than find themselves stacked in standard homes built for them with no input on their part. This process of participation will also advance viability by gradually building bridges between the various returning groups who had been separated from each other for a whole generation.

The Palestinians, as they establish their new state, will have one more unique feature which should be advantageous in creating economic viability. They will have a chance to overcome the traditional hierarchical social structure which has characterized most of the Arab countries for centuries. The Palestinians are now in a position to build a new society which would have a non-traditional leadership, a large number of individually-achieving professionals and entrepreneurs, and the chance to keep it that way. The ability to relate advancement and recognition to achievement, rather than to tradition and family wealth and origin, would be a major step toward creating a dynamic, integrated and viable economy. This will also mean that the duality which has characterized most developing countries will be prevented from reappearing and becoming institutionalized in the new state.

Finally, while it is easy to recognize that industry and manufacturing will eventually emerge as the backbone of the Palestinian economy, it is important in the early stages to encourage crafts and manufactures that have a symbolic and unifying meaning. The new society will have come together in the framework of war, violence, and catastrophe. These 'unifying' cries for the creation of a state will have lost their relevance once a state is established. Therefore, new symbols of unity and co-operation become necessary. National crafts offer good opportunities, not only to supply the tourist industry, but also to sustain the national consciousness of the people and their unity in peace time. New levels of sophistication and technology applied to the national crafts may also be a source of incentive, a measure of capability, and a guide towards building reasonable expectations which can be realized without too much strain on the economy.

With these measures, the new Palestinian state should be expected to meet the subjective criteria of viability. The conditions that presumably will prevail tend to favour the confidence of identity that is basic to economic viability.

Notes

1. Data for this estimation have been extracted from Ben-Shahar *et al. Economic Structure*, p.88.
2. Basic data were taken from A. Bregman, *The Economy of the Administered Areas,* especially p.39; estimates computed by authors.
3. Our estimates are based on data of the United Nations Refugee Office, according to which immovable and movable property lost by the Arabs in Palestine as of 1947-8 would amount to 120 million Palestinian pounds (LP) which were equivalent to sterling pounds. We have assessed the value by compounding that sum for a period of 29 years at a rate of interest of 6 per cent, and converted it into dollars at a rate of exchange of LP 1 = \$4; the total is \$2,600.8 millions; we have raised this to \$3 billions to prevent undue bias of estimate downward. For the basic data see Rony E. Gabbay, A Political Study of the Arab-Jewish Conflict. The Arab Refugee Problem, Geneva: Librairie E. Droz and Paris: Librairie Minard, 1959, pp.342-8; other rough estimate downward. For the basic data see Rony E. Gabbay, *A Political Study of the Arab-Jewish Conflict. The Arab Refugee Problem,* Geneva:
4. There are suggestions that Israel is held together because of its insecurity, rather than by integration and harmony.
5. Here we include the Palestinians who have not acquired citizenship in any country, other than Jordan and Israel.
6. Political and social integration are not treated in this context.
7. Rosemary Sayegh, 'Palestinian Identity: An Experience of Statelessness', *Action,* 13 June 1977, p.8.
8. William T. Bluhm, *Building an Austrian Nation,* New Haven and London:

Yale University Press, 1973, pp.134-5.
9. The Chinese in Indonesia and West Africa, the Turks and Greeks in Cyprus, the Catholics and Protestants in Ireland, and the Blacks in Rhodesia and South Africa, provide examples of failing integration.
10. This does not mean that integration has been complete or fully successful, especially in the US where racial problems are still serious.

8 REGIONAL INTEGRATION

It has been common to suggest that a Palestinian State, to be viable, has to be economically integrated in part or in total with one or more of the countries of the Middle East. We are sceptical about these suggestions. We are actually more inclined to the opposite, namely no integration at all. To be able to reach a conclusion, however, we shall explore the general grounds for integration, and whether integration would benefit the State of Palestine or become a burden to it. We shall treat these issues in that order.[1]

The Meaning and Grounds for Integration

The concept of integration has been used to describe a number of possible collective outcomes, ranging from sectoral co-operation to comprehensive economic unification. Integration has been regarded 'as a process and as a state of affairs'. As a process, 'it encompasses measures designed to abolish discrimination between economic units belonging to different national states'. As a state of affairs, it implies 'the absence of various forms of discrimination between national economies'. The forms it takes in this definition are 'a free trade area, a customs union, and complete economic integration'.[2]

A more comprehensive definition of integration considers society integrated 'if all members of that society are treated equally, enjoy equal opportunities, and an equal degree of liberty, and if they can to the fullest extent achieve whatever goals they may pursue, and in particular more goods and more leisure'.[3]

These definitions are flexible in meaning and application. The flexibility, however, has been regarded as an inherent feature of integration in the sense that as a dynamic process, the conditions for and justification of integration change according to the economic circumstances, the internal contradictions or the dialectic of the process itself. In that sense, integration must be explored within its historical framework, rather than as an abstract general concept.[4]

Integration may be promoted within a country, as well as between countries, and on a sectoral as well as on a regional basis. Whichever form of integration is intended, certain prerequisites must be met to render integration beneficial. These prerequisites may be summarized as the removal of all economic restrictions within the integrated area and a common policy toward the rest of the world. Allowing such a policy to reach its logical conclusion, integration unifies within and restricts without.[5] We are interested in the applied aspects of integration.

Sectoral co-operation or co-operation between members of an industry within a country and in different countries could spell a cartel, which reduces competition between the members and increases their power against the outsiders. Co-operation between the producers of complementary goods could, in contrast, increase competition by giving the producers certain advantages over their competitors who are not in co-operation with the suppliers of their complementary inputs. The same results should be expected when countries, rather than industries, are the actors in the integration process. The outcome of integration, therefore, must be predicated on the conditions of co-operation, the identity of the participants, and the form it takes.

Economic integration would be beneficial if it increases efficiency by lowering relative costs through increased factor and product mobility, economies of scale and externalities, including those which are social and political. Factor and product mobility would be advanced either by removing all barriers to trade, or by 'rational' allocation of resources in accordance with a plan so that bottlenecks may be avoided. The former approach is the remedy prescribed by free trade. However, given the dangers and inequities often entailed by free trade, a co-ordinated effort tends to be the more feasible and reliable approach. Within a country, the national plan is the mechanism for such integration. Between countries, the form is usually bilateral or multilateral agreements which are concluded according to the dictates of the national policy or plan, as practised by socialist countries.

A hybrid and more common approach has been the formation of free trade areas or customs unions among countries within a given region; proximity and contiguity are strong arguments for inclusion in such a union. A free trade area and a less free customs union tend to encourage mobility of inputs and outputs within the designated trade area by favouring the area's products against inports from the outside. This type of integration, therefore, diverts trade in favour of the members.

Whether left up to the market mechanism or promoted by co-ordinated planning, integration would encourage specialization, economies of scale and reduction of waste through redundancy. Thus, the members of the free trade area would move toward more inter-dependence, less self-sufficiency, and more trade between them. The benefits would be maximized if such mobility were extended to all inputs and all outputs, such that the flow of labour, technology and capital would be in total harmony with the market or plan expectations.

The implications of such a high degree of integration are numerous and complicated. For example, economic dislocations would no doubt result in short-run detrimental effects on the weaker industries, and in long-run dependence on the surviving producers. While such effects might offset each other within an economy, they do alter the distribution of economic power within the region, or among the member countries, which may or may not be desirable. Similarly, removal of trade restrictions or co-ordination of the mobility of inputs and outputs should reduce the political and social diseconomies, by lowering the costs of collecting tariffs, or guarding the borders, and possibly of spending less on defence. In other words, economic co-operation would imply less political and military conflict and hence less expenditure on political and military activities. By the same token, participating countries would be less capable of protecting themselves individually, on the assumption that such protection is no longer needed, except against the outside which would be carried out collectively, as NATO is meant to do for the Common Market countries. By implication, then, economic integration is unlikely to succeed because no country would want to suffer the dislocations, reduce its capability to be self-sufficient, or to increase its dependence and vulnerability.

The above analysis has concentrated on mobility of inputs and outputs. Integration, however, may involve more fundamental co-operation by promoting joint production of raw material and finished products, as well as by joint exploitation of natural resources, as in the use of the Nile waters. This type of integration may extend to the joint exploitation of energy, space, etc., but this is more common within nations, in part because the possibilities are limited for joint ventures, but also because it requires a greater commitment or higher investment than integration through trade. Carried to its logical conclusion, integration of national production renders national boundaries insignificant and nationalism merely ceremonial. A country would, therefore, choose to integrate with others, in one form or another, only if the expected benefits exceed the expected costs, economic, political and social. Whether large or small, a country may choose to integrate if it is apparent that the combined net benefits are positive. This means that integration may entail negative economic benefits but confer positive political and social or military benefits such that the trade-off is deemed desirable. The reduction of nationalistic pride may be tolerated for the securement of scarce raw material, or military security may be desired at the expense of economic benefits. While the measurement of these benefits and costs, and their aggregation, may be difficult, the decision to integrate is dependent on the expectation or guesstimate that the net benefit is positive. It is within this framework that we shall assess the desirability of integration between the Palestinian State and its neighbouring countries.[6]

Efficiency and Desirability of Palestinian Integration in the Region

A state of Palestine, as outlined above, may consider integration, in one form or another, with Israel, Jordan, and by extension with Lebanon, Syria and Iraq which are a little removed from the Palestinian territory. Extension of the integration and the form it would take would depend on the initial membership, and the political settlement which accompanies establishment of the State of Palestine. At present, it is our impression that integration is unlikely beyond the three immediate neighbours, Palestine, Jordan and Israel. Therefore, we shall evaluate the costs and benefits and hence the desirability of such integration and the form it may take between these three national entities only.

A look at the natural endowments and the composition of output in the three economic entities shows the following: (1) there is a high degree of similarity between them in endowment; the land, the water, the mines, and the climate are similar. All are short of good land and all face water shortage and enjoy only meager endowments of minerals and raw material. (2) All three depend on the outside for aid, and for direct or indirect investment. (3) All three must depend on industry in order to raise the standard of living in the country — Israel has gone a long way in that direction, which may put Israel in an advantageous as well as a precarious position with regard to integration. (4) There are limited prospects for complementarity among the three economies, on the basis of domestic endowment. (5) Therefore, the most and probably the only advantageous form of integration would be one that promotes division of the market, specialization on the basis of imported inputs, and production by each for a larger market, encompassing the integrated region, and the exchange of these commodities between them. Table 8.1 shows the main exports of the three economies, as indications of specialization and potential complementarity.

The composition of exports and imports shows a high degree of similarity between the three economies, which is not surprising given their proximity to each other and common natural conditions. It is evident that pending major changes in the economic structure of these

Table 8.1: Composition of Exports and Imports of Jordan, Israel and the Palestinian State Territory

	Jordan	Israel	Palestine
Exports	Livestock	Citrus	Citrus
	Cereals & pulses	Fruit	Vegetables
	Seed & feed	Vegetables	Olives
	Foodstuffs	Olives	Fruit
		Flowers	Olive oil
	Copper ore & concentrates	Arms	Soap
	Textile & cotton	Chemicals	Stones
	Paper & printing material	Cut diamonds	Plastic products[a]
	Machines & spare parts[a]	Textiles	Other[a]
	Other (misc.)	Machines & parts[a]	Labour, mostly unskilled
		Rubber tyres	
		Skill (human capital)	
Imports	Durable consumer goods	Durables	Consumer goods — durable & non-durable
	Foodstuffs	Foodstuffs	Agr. inputs[b]
	Machinery	Intermediate goods for agr. & ind.	Machinery[b]
	Finished products	Spare parts & tools	Spare parts[b]
	Arms	Heavy machinery	Other[b]
		Arms & equipment	

a. Re-export.
b. Indirectly through Jordan or Egypt up to 1967 and through Israel since 1967.

economies, they would gain little by integration since they lack the same endowments and can provide few commodities to each other that are not based on imported inputs. This means that as each of these economies develops, producing finished products would be the objective of industrialization and manufacturing. Thus, import substitution would begin by replacing the imports from each other and concentrating on the importation of raw materials and intermediate products which come from outside the region.

However, integration may be significant under one condition: selective industrial specialization based on imported inputs by each of the members of the group. Large-scale industry may be established with a certain division of labour between the different integrated economies to serve all of them as one enlarged market. But this form of integration may be unnecessary if free trade were allowed to function, since competition and free trade would allocate resources according to comparative advantage with full employment. However, in the absence of full employment and in the face of severe competition from outside the integrated area, a conscious policy of integration may be useful. It would make possible the division of the market such that the benefits may be allocated by design, and trade may be diverted in favour of the integrated area. In other words, integration in this case would be a combination of a customs union and planned regional development. The planning would guide the division of labour, and the protective customs union policy would divert trade inwardly. And if full employment is specified as the target, all members of the integrated area would benefit from the integration.

There are dangers in this policy. There is no guarantee that other countries would not retaliate against the union in the exchange of inputs and finished products, or by withholding aid which may be badly needed. This is one of the traditional arguments against customs unions. Strong economies, however, can withstand the pressure, as has been the case with the Common Market. Relatively poor and weak economies may not be able to withstand the pressure, especially if they are dependent on the outside for raw material and intermediate inputs.

The problems of integration, whether bilateral or trilateral, between Palestine and its neighbours are bound to be more internal than external. The differences in the levels of technology and productivity between the prospective member countries render a division of labour more beneficial to the more advanced than to the less advanced, as has been the effect of the division of labour between the developed and the primary producer countries. A free trade area could lead to what has been described as 'Free Trade Imperialism'. An acceptable planned division of labour between unequals would require either full political and social harmony, as might be expected within a national economy, or a high degree of patronage and compromise by the strong member. Such a compromise, however, unless fully justified, could lead to instability and disharmony. This is the kind of relationship that we may anticipate between Palestine, Jordan and Israel if they were to form a free trade or integrated area.

Illustrations of what may be expected already exist in the relationship between the West Bank and Gaza on one side, and Israel on the other. Israel has tried to integrate the economies of these territories with its own. The results have been considered by some people as disastrous. The West Bank and Gaza economies have changed little in their structure in the decade during which they have been under Israeli occupation: industry has experienced little development; the infrastructure has remained stagnant while relatively cheap labour has continued to flow to man Israeli factories and workshops from these territories. As a result, there has been increasing dependence of these economies on Israel, rather than interdependence as would be the result of genuine integration. The difference in levels of technology and the apparent resulting exploitation of the weak by the strong may be seen in the persistent difference in per capita incomes between the West Bank, Gaza Strip and Israel, as shown in Table 8.2.

Table 8.2: Per Capita Income in West Bank, Gaza, Israel (in 1974 prices)

| | IL | | |
	1973	1974	1975
Per capita income/West Bank	2,851	3,527	3,278
Per capita income/Gaza	2,308	2,352	2,333
Per capita income/Israel	13,228	13,104	11,963
Ratio of Israel/West Bank per capita income	4.6:1	3.7:1	3.6:1
Ratio of Israel/Gaza per capita income	5.7:1	5.6:1	5.1:1

Source: Based on Bank of Israel, *Annual Report,* 1975, pp.36-7; Arie Bregman, op.cit., pp.17, 24.

These figures suggest that integration in the sense of promoting more homogeneity, equality and mobility has not taken place, even though some narrowing of the gap has taken place as a result of differential price effects in the two sectors of the 'artificially integrated' economy and a decline in the real per capita income in Israel. Why should these dualistic conditions cease to exist or move in a different direction after the creation of a Palestinian State is not obvious and indeed may be doubtful, given the point of departure, technologically and economically, the Palestinian State will have to start from, compared to Israel.

The structure of industry in the West Bank and Gaza has undergone little change under Israeli occupation. The size of the labour force in industry has remained around 15 per cent of the employed, as late as 1973. Most of the industrial activity has continued to be labour-intensive craft work. Some activities have accelerated, but only to cater to the Israeli market and as extensions of Israeli enterprises, such as in building material and textiles by subcontracting. As Van Arkadie has observed, 'the detailed composition of industrial activity is no more than that typical of the least industrialized primary exporting nations, the largest activities being the food, beverage and tobacco industry and the olive presses. . .for local and export use'.[7] It is equally significant that no new industrial enterprises have been established in the occupied territories, that virtually all industrial imports are received through Israeli intermediaries presumably at higher prices than might have been possible otherwise, given the economic structure of Israel, and that most exports are channelled through Israel.[8] Thus, the main impact of economic co-operation between Israel and the occupied territories has been the growing dependence of the West Bank and Gaza on the economy of Israel, in the tradition of colonial economic relations or imperialism of the past century and a half. To quote Van Arkadie again, 'two major characteristics define the economic relationship that has developed between Israel and the West Bank and the Gaza Strip — the growth of a market for local labor in Israel, and the growth of a market for Israeli commodities in the two territories. Quantitatively, these developments outweigh all other effects, which at least in the short run have been comparatively minor.'[9]

Similar differential effects were experienced between 1948 and 1967 when the West Bank was a part of Jordan. While mobility of inputs and outputs was relatively free, administered investment was encouraged more on the East Bank than on the West Bank, especially in industry and manufacturing. For example, of fifteen companies existing in 1967 in which the government had a share, only four were on the West Bank. The government of Jordan placed only 8 per cent of its investment in large companies in the West Bank; the rest was invested in the East Bank. And when unemployment struck, it hit the West Bank first and more seriously than it hit the East Bank. The effects may also be seen in the wage differentials. The average wage in Jordan in 1965 was 176 Jordanian dinars (JD). In Amman district the average wage was JD 221, compared with JD 141 in the Jerusalem district, JD 113 in Nablus district and JD 151 in the Hebron district.[10]

The idea of integration between these parties is not new, although it has often taken a different character. The Palestinians have always wanted a form of integration with Israel in what they describe as a unified, secular, democratic state which, if implemented, would eventually render the Jews a minority and change the character of the

Israeli entity. Similarly, as an alternative to continued occupation, Israel has been insisting on some form of integration of the Palestinians and the West Bank territory into Jordan as the solution. If carried out, that solution would render the Palestinian entity formally non-existent. Thus, either type of integration would mean annihilation in the political sense. And if either of these parties were made economically dependent on the other(s) before a high degree of trust and harmony had been created and before a certain degree of economic competitiveness had been achieved, domination and figurative annihilation would certainly follow. While regional viability may be thus promoted, the viability of some of the units in the region will be seriously damaged.

The logical conclusion to this analysis, if the respective national economies are to be preserved, is to have no integration and to have no completely free trade. Guided trade and specialization may be the extent of beneficial integration, which is little different from the policies that have prevailed in many regions of the world. Total integration, however, need not be the only possible way of co-operation.

Integration among the three national economies, Palestine, Jordan and Israel, may be promoted on a more controlled but also more beneficial level than may be expected from the types of integration discussed above. Given the limited resources in the region, the proximity of these resources to each other and the ease with which they can be pooled together for exploitation, it should be feasible and highly beneficial to integrate on a sectoral or project by project basis, with one or both of the other parties. Let us call this *Project Integration* and see why it should be highly recommended.

1. Project Integration may be politically neutral and consistent with national sovereignty and economic independence since it is primarily a joint investment, which may be undertaken only by full and voluntary agreement by the respective parties.

2. Project Integration allows flexibility such that joint projects may be carried out only if the prospective benefits have been found positive. Failure to participate in a project need not endanger co-operation in other projects.

3. Product Integration promotes equity and fairness in the distribution of benefits and spillovers, economically, technologically and even socially. The joint exploitation of a resource brings together human resources that are of equal significance to the project. It promotes communication between the experts and staff. It guarantees supply of the project to the member countries, and it could promote harmony and understanding on a wider scale in the process.

4. Project Integration may be experimental, beginning on a narrow scale and at a relatively low technological level. This process would minimize the cost of failure and enhance the chances of success in future integrated projects. In other words, Project Integration permits the parties to evolve toward more comprehensive integration as they bcome ready for it.

5. Project Integration may be efficiency oriented: it permits complementarity, specialization and scale production and marketing.

Several industries may be candidates for Project Integration. Water exploitation of the Jordan and the Yarmuk is the classic example. The Johnston Plan proposed during the Eisenhower presidency was too politically oriented and therefore it was choked in infancy. Once a Palestinian State has been established, a different atmosphere will prevail, which may be more conducive to the project implementation. Water desalination is another potential candidate. Studies by Jerome Fried for the joint revival of the desert through desalination are good examples.[11] Potash and phosphate exploitation in the area of the Dead Sea is still another possible project.

Project Integration has an advantage over other types of integration in still another way, especially for the State of Palestine: it gives the state enough time to establish its national identity securely and freely. The State of Palestine will need time to feel secure enough to co-operate with the parties that had inflicted loss and pain on its people in the recent past. A new perception must evolve. The confidence of identity must be established and made permanent, without the threat of economic exploitation or dependency through some form that allows this freedom, which applies equally to the other parties involved in integration.

Given these arguments, the following are our recommendations:

1. Federation, confederation, or any such form of economic (or political) integration between Palestine, Jordan, and Israel on a national level poses more costs than benefits to the viability of the Palestinian state, at least in the foreseeable future.

2. A free trade area or customs union between these parties is potentially dangerous as it may promote 'free trade imperialism' among them and evoke retaliation against them from the outside.

3. Until the confidence of identity by the state of Palestine has been achieved, and until a certain degree of economic equality and competitiveness between the parties has been realized, Project Integration is the only form of economic integration that seems feasible, harmless and economically beneficial to all the parties.

Notes

1. We distinguished in our analysis federation from confederation. To the extent that confederation may be decided upon by mutual agreement, and to the extent to which it guarantees autonomy of the individual confederate units and precludes exploitation and domination by other members, we are not opposed to confederation and may even support it.
2. Bela Balassa, *The Theory of Economic Integration*, Homewood, Ill.: Irwin, 1961, pp.1-2.
3. Paul Streeten, *Economic Integration*, Leyden: A.W. Sythof, 1961, p.17.
4. B.N. Ganguli, *Integration of International Economic Relations*, London: Asia Publishing House, 1968, Ch. 1.
5. A detailed list of minimum conditions can be found in Victoria Curzon, *The Essentials of Economic Integration*, London: Macmillan, 1974, pp.19-25.
6. Integration with other and large and powerful countries may be desirable, but in this context it is rather irrelevant; e.g., integration with the US, France, or the USSR would confer such benefits that it would be hard to reject, except on the basis of nationalistic pride.
7. Bran Van Arkadie, *Benefits and Burdens. A report on the West Bank and Gaza Strip Economies since 1967*, Carnegie Foundation, 1977, p.124.
8. Ibid., pp.88 ff.
9. Ibid., p.141.
10. Jamil Hilal, *The West Bank*, pp.132-144.
11. *Desalting Technology and Middle Eastern Agriculture*, Praeger, 1971; other studies have been circulated more recently by the Middle East Institute, Washington, DC, in which Fried is an Associate.

9 SUMMARY AND CONCLUSIONS

Hypotheses and Findings

Two main objectives have guided this study: to explore and clarify the meaning and conditions of economic viability of small nation-states, and to assess the prospects of economic viability of an Arab State of Palestine within the pre-1967 war boundaries. We have approached these objectives by presenting certain hypotheses regarding the role of endowment, size and economic structure. We also proposed that a State of Palestine on the West Bank and Gaza would be economically viable.

To test these hypotheses, we have used comparative analysis. To deal with the concept of viability we have looked at various countries in different regions, and with different types of endowment. The economic viability of the Palestinian State has been assessed by setting up viability targets and exploring the degree to which these targets can be met. In the process we have surveyed the resources and human factor of the prospective state, the required capital for its development, and the potential sources of capital. We have also investigated the benefits and costs of integrating the Palestinian State with one or more of its neighbouring countries. The data we have used have come mainly from official sources. However, these data were limited because of the hypothetical nature of the proposed state. Even though we have tried to deal with the issues rigorously, the conclusions reached must be treated with great care, pending further analysis and study as proposed below. Our findings may be summarized as follows:

1. Economic viability may be achieved if the given country can attain and sustain a per capita income comparable with those of other countries in the region with comparable resources. To be able to achieve and maintain viability, a country must be able to maintain a balance between its population and other resources. Hence, the size of the nation whether in terms of population, territory or endowment may have little significance by itself. As long as a minimum of the irreproducible resources is available, a congenial climate prevails, and population exists, viability is potentially achievable. Given the prospective population, a certain amount of land and water is necessary; with a climate suitable for human settlement, the critical condition for viability is the ability to maintain a balance between the population and other resources. The reproducible resources may be

obtained from outside the country. However, to be able to achieve and maintain viability, there is a subjective factor that is critical. This is the human factor which includes the ability to identify reasonable expectations, to recruit resources, to assess and respond to the demand, and to pursue economic goals with enterprise and efficiency. We have summarized this human factor as the confidence of identity for the nation: without the confidence of identity, the prospects for viability may be poor or non-existent.

2. A State of Palestine on the West Bank and Gaza (including East Jerusalem) can be economically viable. Viability, however is not inherent; it has to be achieved. We estimate that the State of Palestine can support far more than the two and a half million people expected in the first few years. The State of Palestine will have all the critical irreproducible resources, should be able to secure the reproducible resources, and we expect the Palestinians to be able to realize fully their confidence of identity if the chance prevails. The State of Palestine will have to depend heavily on urban industry. The shift to industry will be co-ordinated with the eventual decline of the building construction that may come about as all the returning and homeless refugees have been resettled. Agriculture will play a relatively declining role unless a major \ breakthrough in desalination takes place.

3. The irreproducible resources are ample to permit viability. The reproducible resources will amount to less than $12 billions at the 1977 prices (estimated) for the whole first five years which may be considered the transition period. Though it may be hard to specify the sources of capital at this point, we have charted the potential sources as an illustration of ways to secure the capital. We have also charted the employment structure in the new economy, and estimated the potential national income during each of the first five years. The expected income falls short of the target income by less than $300 million a year; however, if we allow for annual growth and for higher productivity in industry as new capital is invested, the gap between the potential and target incomes will vanish.

4. A State of Palestine would be more likely to achieve its confidence of identity and economic viability if it stays independent and refrains from integration with its neighbours. Integration with Israel and/or Jordan carries the risk of economic imperialism by one party or another, and a threat to the confidence of identity of the new state of Palestine. After the state had achieved viability, large-scale integration with other economies may be considered. In the meantime project integration may be the limit of harmless interdependence with

other countries.

Pending Issues

A number of questions relating to implementation remain to be researched. Some of these are long-term projects, while others are politically sensitive and can be approached only after the political conflict has been reduced considerably. Among these issues are the following:

1. A detailed survey and occupation profile of the prospective manpower is badly needed. We have no data on how many engineers there are, how many mechanics, etc., what specializations are available and how many qualified people may be expected to return to the State of Palestine. What is the rate of labour absorption that may have to be planned? Such a survey may be difficult to conduct at this time, unless it is sponsored by the Palestinian leadership.

2. Since viability requires a balance between population and other resources, a study of population growth or more specifically of population control among the Palestinians is essential. Will the Palestinians be able to control the population growth and make it feasible to maintain a balance with the resources without lowering the standard of living of the people?

3. Given the fragmented layout of the land and the traditional land tenure system in the West Bank and Gaza, agrarian reform in the new state may be indispensable to make the best use of the available land area. Little attention has been given to this question as yet, even though reform of the land tenure system may be most appropriate and efficient if carried out as the state is being established. A plan of agrarian reform, including goals and processes of reform, deserves to be formulated and ready even as a political settlement is being signed.

4. It is necessary to study the prospective industries in detail in order to identify the feasible and promising industries for the state. These industries can be identified only after the economic system and institutions have been outlined and the marketing channels have been explored. What is the source of demand for the new products, where would the raw material come from, how much capital would be needed for each? These are some of the questions that need to be answered.

5. Though we have made suggestions regarding the sources of capital, a thorough assessment of these sources remains to be undertaken. It may be appropriate even to seek commitments from the respective parties. For example, how much can be expected from the Arab countries and in what form will it be dispensed? How much compen-

sation from Israel should be expected and at what rate may these funds be expected to flow? What financial institutions must be created to cope with the economic activity that will be generated by the creation of a new state? Given the poverty of the infrastructure and financial institutions in the West Bank and Gaza at the present time, how will the functions of these institutions be satisfied?

6. Finally, implementation means a plan of action. Assuming a State of Palestine is to be created as outlined, how will repatriation of the Palestinians take place? They cannot all come at the same time; how are priorities to be established as to who may return and get settled first? What methods of resettlement and rehabilitation will be followed? Will these be carried out by state agencies or will they be left largely to the market? These questions may have been addressed by the Palestinian authorities but we have no documentation of that. Yet most of these questions cannot be addressed adequately without the full co-operation of the Palestinians and the countries they reside in, and not before the political and geographical boundaries of the state have been demarcated.

It is our hope that this study has answered the general question of the economic viability of the state of Palestine, and that it has rendered the study of the pending issues a logical next step in the direction of solving the Arab-Israeli conflict.

APPENDIX

Table A1: The Output and Investment per Employee (in $) According to Commodities

Group of Commodities	Output per Employee in $	Investment per Employee in $
1. Bakery and noodle products	6,700	15,000
2. Canning and preserving of fish, fruit and vegetables	6,750	2,700
3. Chocolate and sweets	8,900	15,600
4. Tobacco products	44,200	50,000
5. Pharmaceutical industry	8,350	3,700
6. Cosmetics and perfumery	15,800	10,300
7. Soaps, detergents and similar products	12,900	8,800
8. Rubber products (excl. tyres)	11,500	13,700
9. Synthetic and wool yarns	9,400	12,000
10. Spinning cotton gins	6,750	11,050
11. Weaving of fabrics & manufacture of synthetic yarns and fabrics	9,100	9,880
12. Knitting mills	6,750	9,500
13. Embroidery	1,750	500
14. Tents, water courses, tarpaulin	8,700	1,200
15. Blankets	7,800	980
16. Woven & knotted carpets	8,700	1,200
17. Cement	7,800	41,700
18. Clay and lime products		
19. Tools, pipes and fittings,	3,000	50,000
of iron or steel	4,750	5,780
20. Nails, screws, etc., iron, steel or copper	3,980	7,160
21. Tools for manual use	4,080	3,770
22. Cutlery	1,400	2,170
23. Heating and cooking equipment and utensils	5,500	1,500
24. Agricultural machinery	7,700	2,200
25. Commercial and domestic machinery	9,600	2,400
26. Electrical motors and transformers	3,680	5,930
27. Apparatus for electrical circuits; telecommunications apparatus	8,800	10,200
28. Electrical supplies (incl. batteries and accumulators)	17,000	5,100
29. Manufacture of cars, motorcycles and parts	18,500	9,000
30. Carpentries for furniture and building	4,160	2,060
31. Outwear (incl. tailors and dressmakers)	5,900	1,100
32. Leather products	5,000	800
33. Underwear (other than knitted)	4,060	2,830
34. Manufacture and repair of footwear	11,200	4,950
35. Plastic products	12,000	10,200

Table A2: List of Products and Findings of Import Forecasts for Investigated Markets (total import in thousand dollars in 1968)

Group of Products	OECD countries	Middle East	Israel
1. Bakery and noodle produce	137,105	3,113	—
2. Canning and preserving of fish, fruit and vegetables	617,891	2,546	34
3. Chocolate and sweets	187,441	3,236	166
4. Tobacco products	—	36,873	1,773
5. Pharmaceutical industry	—	96,938	5,892
6. Cosmetics and perfumery	144,383	14,239	693
7. Soaps, detergents and similar products	233,159	4,365	210
8. Rubber products (excl. tires)	577,233	50,998	1,994
9. Synthetic and wool yarns	1,755,739	62,059	15,895
10. Spinning cotton gins	134,243	1,276	—
11. Weaving of fabrics and manufacture of synthetic yarns and fabrics	760,288	26,170	780
12. Knitting mills	—	8,642	864
13. Embroidery	—	7,050	200
14. Tents, water courses, tarpaulin	23,276	—	—
15. Blankets	—	8,600	28
16. Woven and knotted carpets	506,588	10,481	106
17. Cement	—	11,076	415
18. Clay and lime products, Fire resistant blocks.	130,000	—	—
19. Tools, pipes and fittings of iron or steel	153,579	—	—
20. Nails, screws, etc. iron, steel, or copper.	345,922	—	—
21. Tools for manual use	505,267	29,800	3,350
22. Cutlery	182,282	11,565	454
23. Heating and cooking equipment and utensils (non-electric)	244,681	—	—
24. Agricultural machinery	1,242,004	—	—
25. Commerical and domestic machinery	—	8,410	1,527
26. Electrical motors and transformers	—	61,418	11,446
27. Apparatus for electrical circuits; telecommunications apparatus	1,090,353	—	—
28. Electrical supplies (incl. batteries and accumulators)	119,153	5,838	717
29. Manufacture of cars, motorcycles and parts	—	73,208	10,302
30. Carpentries for furniture and building (incl. upholstery)	699,130	16,600	1,595
31. Outwear (incl. tailors and dressmakers)	1,308,665	17,814	290
32. Leather products	121,301	658	73
33. Underwear (other than knitted)	1,310,901	17,556	2,610
34. Manufacture and repair of footwear	1,293,602	7,914	1,000
35. Plastic products	560,271	9,971	2,294

BIBLIOGRAPHY

Books

Abu-Lughod, I. (ed.). *The Transformation of Palestine,* Evanston: Northwestern University Press, 1971.

Balassa, Bela. *The Theory of Economic Integration,* Homewood, Illinois: Richard D. Irwin, Inc., 1961.

——, *Economic Development and Integration,* Centro De Estudios Monetarios Latinoamericanos, Mexico, 1965.

Ben-Shahar, H. Eitan Berglas, Yair Mundlak and Ezra Sadan. *Economic Structure and Development Prospects of the West Bank and Gaza Strip,* Santa Monica. Rand, 1971.

Bluhm, W.T. *Building an Austrian Nation,* New Haven: Yale University Press, 1973.

Boulding, Kenneth. *Conflict and Defense,* Harper Torchbook, 1963.

Bregman, Arie. *The Economy of the Administered Areas,* Jerusalem: Bank of Israel, 1976.

Buehrig, E.H. *The UN and the Palestinian Refugees: A Study in Non-territorial Administration,* Bloomington and London: Indiana University Press, 1971.

Bull, Vivian. *The West Bank – Is It Viable?* Lexington Books, 1976.

Cattan, H. *Palestine, the Arabs and Israel,* Longmans, 1969.

Chenery, H. and M. Syroquin. *Patterns of Development, 1950-1970,* Oxford University Press, 1975.

Clawson, Marion (ed.). *Natural Resources and International Development,* Baltimore: Johns Hopkins University Press, 1964.

Cohen, Abraham. *The Economy of the Territories,* Ayn-Hahorish, 1975 (Hebrew).

Curzon, Victoria. *The Essentials of Economic Integration,* London: Macmillan, 1974.

Darin-Drabkin, H. *Urban Land Policies,* United Nations, 1973.

——. Land Policy and Urban Growth, forthcoming.

Davis, Uri, Andrew Monck and Nira Yuval-Davis. *Israel and the Palestinians,* London: Ithaca Press, 1975.

Demas, W.G. *The Economics of Development in Small Countries with Special Reference to the Caribbean,* Montreal: McGill University Press, 1965.

Fabricant, Solomon, 'Perspective on the Capital Requirements

Question', in Eli Shapiro and W.L. White (eds). *Capital For Productivity and Jobs,* Prentice-Hall, 1977.

Fried, Jerome F. and M.C. Edlund. *Desalting Technology and Middle Eastern Agriculture,* Englewood Cliffs, N.J.: Praeger, 1971.

Fried, Jerome F. *A North Sinai-Gaza Development Project,* The Middle East Institute, Washington, D.C., 1975 (mimeograph).

Gabbay, E.R. *A Political Study of the Arab-Jewish Conflict. The Arab Refugee Problem.* Geneva: Librairie E. Droz and Paris: Librairie Minard, 1959.

Ganguli, B.N. *Integration of International Economic Relations,* London: Asia Publishing House, 1968.

Hadawi, Sami. *Bitter Harvest: Palestine Between 1914-1967,* N.Y.: The New World Press, 1967.

Hilal, Jamil. *The West Bank. Social and Economic Structure, 1948-74,* PLO Research Centre, Beirut: Palestinian Book Series L60, 1974 (Arabic).

Jiryis, S. 'The Legal Structure for the Expropriation and Absorption of Arab Lands in Israel', *Journal of Palestine Studies,* vol.II, no.4 (Summer 1973), pp.82-104.

Johr, W.A. and F. Kneschaurek. 'A Study of the Efficiency of a Small Nation: Switzerland', in E.A.G. Robinson (ed.), *Economic Consequences of the Size of Nations.*

Kanovsky, E. *The Economic Impact of the Six Day War,* N.Y.: Praeger, 1970.

Khalaf, N.G. *Economic Implications of the Size of Nations, with Special Reference to Lebanon,* Leiden, 1971.

Knorr, Klaus. *Power and Wealth,* N.Y.: Basic Books, Inc., 1973.

Kohr, Leopold. *Development without Aid. The Translucent Society,* Llandybie, Carmarthenshire: The Merlin Press, 1973.

Lewis, W.A. *Development Planning,* N.Y.: Harper and Row, 1966.

Orgler, Y. *Establishment of Industrial Towns Near the Green Line,* Tel Aviv: Cherikover, 1973.

Peretz, Don. *A Palestinian Entity,* Washington, D.C.: The Middle East Institute, 1970.

Prachoawy, Martin F.J. *Small Open Economies,* Lexington, Mass: Lexington Books, 1975.

Rikhye, I.J. and John Volkmar. *The Middle East and the New Realism,* N.Y.: International Peace Academy, 1975.

Robinson, E.A.G. (ed.). *Economic Consequences of the Size of Nations,* N.Y.: St Martin's Press, 1973.

Rostow, W.W. *The Stages of Economic Growth,* Cambridge, England:

Cambridge University Press, 1960.

Shapiro, Eli and William L. White. *Capital for Productivity and Jobs,* Englewood Cliffs, N.J.: Prentice-Hall, Inc., 1977.

Sheehan, E.R.F. *The Arabs, Israelis, and Kissinger,* N.Y.: Reader's Digest Press, 1976.

Sid-Ahmed, M. *After the Guns Fall Silent,* London: Croom Helm, 1976.

Streeten, Paul. *Economic Integration,* Council of Europe, European Aspects, Series B: Economics, No.5, 1961.

Svennilson, I. 'The Concept of the Nation and its Relevance to Economic Analysis', in E.A.G. Robinson (ed.), *The Economic Consequences of the Size of Nations,* 1973.

Tinbergen, Jan. *International Economic Integration,* revised edition. Amsterdam-London-New York: Elsevier Publishing Company, 1965.

Trinidad, Office of the Premier and Ministry of Finance, *The Economics of Nationhood,* Trinidad, 1959.

Tuma, Elias. 'International Interdependence and World Welfare' in N.M. Kamrany, (ed.), *The New Economics of the Less Developed Countries: Changing Perceptions in the North-South Bargaining,* Colorado: Westview Press, forthcoming.

Vakil, C.N. and P.R. Brahmanada. 'The Problems of Developing Countries,' in E.A.G. Robinson (ed.) *The Economic Consequences of the Size of Nations,* 1973.

Van Arkadie, Brian. *Benefits and Burdens: A Report on the West Bank and the Gaza Strip Economies since 1967,* N.Y. and Washington, D.C.: Carnegie Endowment for International Peace, 1977.

Vinelle, D. de la. 'Study of the Efficiency of a Small Nation: Belgium,' in E.A.G. Robinson (ed.), *The Economic Consequences of the Size of Nations,* 1973.

Ward, R.C. Don Peretz and Evan M. Wilson. *The Palestine State,* Port Washington, N.Y.: Kennikat Press, 1977.

Periodicals and Newspaper Articles

Bulletin of Peace Proposals, Special Issue. 'The Arab-Israeli Conflict', Oslo, 1976, no.4.

New Outlook, various issues.

Preuss, Teddy. 'Palestine in the West Bank-State or Big Labour Camp?' *Davar,* 12 April 1974.

Sayegh, R. 'Palestinian Identity: An Experience of Statelessness', *Action,* 13 June 1977.

Sayigh, Y.A., 'Palestinian Peace', *The Middle East Newsletter,* vol.4, nos.4 and 5, June-July 1970,

Sheehan, E.R.F. 'A Proposal for a Palestinian State', *New York Times Magazine,* 30 January 1977.

Tuma, Elias H. 'Divide Area Fairly Between Two States', *Los Angeles Times,* 1 December 1974.

Documents

Bank of Israel, *Annual Reports, 1974 & 1975.*

State of Israel, *Statistical Abstracts.*

UN General Assembly, *Report of the Commissioner-General of the United Nations Relief and Works Agency for Palestine Refugees in the Near East,* A/10013, 1975.

INDEX

Africa 29
agrarian reform 77, 116
agricultural development 60; marketing 63
Amman 110
Arab fund for Palestine 90, 95
Arab-Israeli conflict 13, 32, 34
Asia 29
Austria 98
Autarky 13

balance of trade 68
Beit Jala 58
Belgium 29
Ben Shahar 36, 37
Bethlehem 54, 57
Boulding, K. 17
Boundaries 70; pre-1967 43-4
building materials 57
Bull, Vivian 39

camp Palestinians 71, 97
capital: markets 28; requirements 94; sources 90, 116
Caribbean 28
Cartels 104
Chenery, H. 27
China 95
Common Market 105, 108
construction and GNP 66, 83
crafts and integration 101
crops 62, 64
Customs Union 104, 108
Cyprus 96

Dead Sea 53, 57, 58, 111; development plan 58; rainfall 74
Dead Sea Coast 53
Demas, W.G. 27
dualism 27, 100, 109

East Bank 36, 37, 38, 110
East Dead Sea 58
East Jerusalem 36, 37, 43, 47, 57, 115
East Valley 55
education 36, 37, 49, 84
Egypt 71
Ein Feshka 58
Eisenhower 112

emigration 47, 51
employment: in agriculture 51, 60, 77, 89; construction 51, 84; industry 90; services 51; of Palestinians in Israel 51; projections 91
exports 63, 82; composition of 107

family structure 49
fertilizer use 63
foreign aid 38, 94
France 95
free trade imperialism 108
Fried, Jerome 112

Gaza 39, 43, 44, 49, 53
Gaza: geography of 58; rainfall 76; refugees in 47, 58, 59, 67
Germany 95

Haifa 70
Hebron 53, 55, 56, 57, 75, 78, 110
Hilal, Jamil 38
hoarding 69

imperialism 38, 115
import substitution 28
industry 37, 64-5; and integration 100; capital requirements 82; feasible 106
industrial towns 81-2
integration 26, 42, 115; and free trade 108, 111; definitions 103; economic 90, and nationalism 105; regional 31, 103, 110-111; within the region 31, 96, 97, 99
interdependence 31, 115
infrastructure 25, 27, 84, 117
Iraq 106
Ireland 85, 90, 96
irrigation 60, 62
isolationalism 26
Israel 13, 29, 33, 36, 37, 38, 39, 41, 43, 44, 47, 48, 58, 60, 68, 69, 70, 90, 96, 106, 108, 111, 115; compensations 117; water consumption 75

Japan 95

Jenin 53, 54, 55
Jerusalem 39, 40, 43, 44, 53, 55, 57,
    58, 75, 110; as an open city 43,
    44; tourism 38
Jewish Zionism 41
Johnston Plan 112
Jordan 39, 40, 43, 48, 58, 60, 70, 71,
    106, 108, 111, 112, 115
Jordan River 13, 15, 33
Jordan Valley 53, 58, 74, 77; water
    consumption 77
Jordanian Educational Act 49

Khalaf, N.G. 27
Khan-Yunis 58, 59
Kirmizan 58
Kuwait 96
Kuznets, Simon 19

Labour force participation rate 50,
    89; female 51
Labour productivity 63; skilled 37
land consolidation 77; cultivation 54;
    fragmentation 64, 77; tenure 77,
    116; use 59
Latin America 29
Lebanon 29, 70, 71, 96, 106

Malthusian trap 31
Mandate, Palestine 42
manufacturing 37
Mar Scibba 58
Mediterranean 38; coast 58; plain 53
minerals 57, 112
Mount Hordua 58

Nablus 53, 54, 55, 57, 58, 110
National accounts: identity 41
NATO 105
Nazareth 70
Nebi Mussa 58
Neper 76, 94
Netherlands 78
Nile 105
Norway 85

open bridges 38

Palestine geography 52-3
Palestinian: armed forces 73; region
    39; state 13, 15, 33, 34, 38, 43,
    44, 46, and manpower 37, 72, as
    Labour camp 40, economic
    viability 41, 95, income per
    capita 46, projections 90, problems

of 84
Palestinianism 98
Palestinians in Israel 71
Partition Plan 42
pesticide use 63
planning 14, 26, 97, 100; and
    integration 104
PLO 35-6, 41, 98
political settlement 38
population: age distribution 50;
    density per room in West Bank
    and Gaza 49; distribution 54;
    forecasts 73, 90; growth 89, 116;
    ratio of males to females 50; rural
    57
Preuss, Teddy 39, 40, 41
Productivity: in agriculture 51; in
    construction 51
Project Integration 111-12

Quman 58

Rafah 76
rainfall 52, 74
Ramallah 53, 54, 55, 57, 74, 75
Rand study 39
refugees 39, 58, 70, 86, 94, 100;
    return of 37, 70, 71, 83, and
    integration 100
remittances 67
Rostow, W.W. 27
rural settlements 57

Sartala 58
Sayigh, Yusuf 36
Sebastia 58
self-sufficiency 14
services 37
Sheeham, Edward 44
Sinai 41
soil categories 55
struggle identity 97
surplus labour 27
Switzerland 29
Syria 71, 96, 106

take-off 27, 28
technology 23
Tel Balata 58
tourism 37, 40, 58, 66, 67, 84
tractors 63
trade diversion 104
transition period 89, 94
Trinidad 28; and Tobago 29
Tulkarm 53, 54, 55

Turkey 85, 90
two-state solution 42

UK 85, 95
UN 30, 41, 43
UNRWA 48, 49, 67, 70, 94; and
    social services 71, 96
unproductive lands 58
urban industry 115; land require-
    ments 80; population and space
    57; space use 79, 80
USA 85, 95
USSR 95

Van Arkadie 110
veterinary services 64
viability: and boundaries 15; and
    capital 94; and land 24-5; and
    national identity 20; and per
    capita GNP 20; and resources 22;
    and take-off 27; as function of
    14; conditional 17; definitions 17;
    dynamic 25; economic 14;
    prerequisites 114; impact of
    culture and tradition on 19; of
    Palestinian State 35; political
    and social 14; regional 111; success
    requirements 19

Vinelle, D. de la 29
vulnerability 18

Wadi Kelt 58
Wadis 53, 58
Wakf 56
Ward, Richard 38, 39
Water capacity 75; consumption
    75; desalination 76; supply
    37, 64, 74
West Bank 48, 60, 82, 83, 108;
    GNP of 38; under Jordan 109;
    water consumption 75
West Bank and Gaza: agricultural
    output 62; costs of contact 40;
    housing 85; economic prospects
    36; employment: agricultural
    37, construction 66, 83,
    industrial 37; manufacturing
    as per cent of GNP   65,
    per capita GNP 67 ;
    refugee density in 50;
    under Israel 65, 109

Yarmuk 112